SUSTAINABLE SOLAR ENERGY SYSTEMS

Challenges and Economics for the Arab World

AF885143

Salah El-Haggar

Department of Mechanical Engineering
The American University of Cairo
Egypt

Sarah Mousa

Arab Studies and Development
Founder of Shamsina, Wynantskill, New York
USA

&

Mohamed El-Morsi

Department of Mechanical Engineering
The American University of Cairo, Ain Shams University
Egypt

BENTHAM SCIENCE PUBLISHERS LTD.

End User License Agreement (for non-institutional, personal use)

This is an agreement between you and Bentham Science Publishers Ltd. Please read this License Agreement carefully before using the ebook/echapter/ejournal (**"Work"**). Your use of the Work constitutes your agreement to the terms and conditions set forth in this License Agreement. If you do not agree to these terms and conditions then you should not use the Work.

Bentham Science Publishers agrees to grant you a non-exclusive, non-transferable limited license to use the Work subject to and in accordance with the following terms and conditions. This License Agreement is for non-library, personal use only. For a library / institutional / multi user license in respect of the Work, please contact: permission@benthamscience.org.

Usage Rules:

1. All rights reserved: The Work is the subject of copyright and Bentham Science Publishers either owns the Work (and the copyright in it) or is licensed to distribute the Work. You shall not copy, reproduce, modify, remove, delete, augment, add to, publish, transmit, sell, resell, create derivative works from, or in any way exploit the Work or make the Work available for others to do any of the same, in any form or by any means, in whole or in part, in each case without the prior written permission of Bentham Science Publishers, unless stated otherwise in this License Agreement.
2. You may download a copy of the Work on one occasion to one personal computer (including tablet, laptop, desktop, or other such devices). You may make one back-up copy of the Work to avoid losing it. The following DRM (Digital Rights Management) policy may also be applicable to the Work at Bentham Science Publishers' election, acting in its sole discretion:

- 25 'copy' commands can be executed every 7 days in respect of the Work. The text selected for copying cannot extend to more than a single page. Each time a text 'copy' command is executed, irrespective of whether the text selection is made from within one page or from separate pages, it will be considered as a separate / individual 'copy' command.
- 25 pages only from the Work can be printed every 7 days.

3. The unauthorised use or distribution of copyrighted or other proprietary content is illegal and could subject you to liability for substantial money damages. You will be liable for any damage resulting from your misuse of the Work or any violation of this License Agreement, including any infringement by you of copyrights or proprietary rights.

Disclaimer:

Bentham Science Publishers does not guarantee that the information in the Work is error-free, or warrant that it will meet your requirements or that access to the Work will be uninterrupted or error-free. The Work is provided "as is" without warranty of any kind, either express or implied or statutory, including, without limitation, implied warranties of merchantability and fitness for a particular purpose. The entire risk as to the results and performance of the Work is assumed by you. No responsibility is assumed by Bentham Science Publishers, its staff, editors and/or authors for any injury and/or damage to persons or property as a matter of products liability, negligence or otherwise, or from any use or operation of any methods, products instruction, advertisements or ideas contained in the Work.

Limitation of Liability:

In no event will Bentham Science Publishers, its staff, editors and/or authors, be liable for any damages, including, without limitation, special, incidental and/or consequential damages and/or damages for lost data and/or profits arising out of (whether directly or indirectly) the use or inability to use the Work. The entire liability of Bentham Science Publishers shall be limited to the amount actually paid by you for the Work.

General:

1. Any dispute or claim arising out of or in connection with this License Agreement or the Work (including non-contractual disputes or claims) will be governed by and construed in accordance with the laws of the U.A.E. as applied in the Emirate of Dubai. Each party agrees that the courts of the Emirate of Dubai shall have exclusive jurisdiction to settle any dispute or claim arising out of or in connection with this License Agreement or the Work (including non-contractual disputes or claims).
2. Your rights under this License Agreement will automatically terminate without notice and without the need for a court order if at any point you breach any terms of this License Agreement. In no event will any delay or failure by Bentham Science Publishers in enforcing your compliance with this License Agreement constitute a waiver of any of its rights.
3. You acknowledge that you have read this License Agreement, and agree to be bound by its terms and conditions. To the extent that any other terms and conditions presented on any website of Bentham Science Publishers conflict with, or are inconsistent with, the terms and conditions set out in this License Agreement, you acknowledge that the terms and conditions set out in this License Agreement shall prevail.

Bentham Science Publishers Ltd.
Executive Suite Y - 2
PO Box 7917, Saif Zone
Sharjah, U.A.E.
Email: subscriptions@benthamscience.org

BENTHAM
SCIENCE

CONTENTS

FOREWORD

Egypt and the Arab world are blessed with abundant solar energy resources. Located in the "sun belt" with powerful solar radiation and dependably clear skies, and having vast tracts of vacant desert land suitable for mounting solar installations, this region is the ideal venue for adoption of solar technology. A formidable challenge facing the Arab world, however, is to harness the seemingly limitless potential of solar energy as a practical engine for social and economic development. It is a challenge that Dr. Salah El-Haggar and the coauthors of Sustainable Solar Energy Systems meet head on.

This volume goes beyond conventional studies of solar technology and economics to address the everyday needs of underserved populations, particularly the rural poor. As Dr. El-Haggar and his co-authors vividly describe, even when fossil fuels (*e.g.* kerosene, diesel, wood) are available to such groups and are affordable, dependency on conventional fuels for household needs and agricultural work undermines their health, saps educational opportunity and blocks economic advancement. This is especially true for women and children.

The authors first provide the reader with a basic grounding in the science of solar energy, both thermal and photovoltaic, the engineering of solar energy systems and the economics of typical solar energy applications. As the authors point out, the world is in the midst of a veritable "solar revolution," with steadily falling prices for solar panels and continuous improvements in solar technologies.

Once past these basics, Dr. El-Haggar and his colleagues demonstrate how solar energy can readily be adapted for rural household and agricultural needs such as cooking, domestic hot water, heating, cooling and ventilation of houses, drying of agricultural products and off-th--grid electricity requirements. Of special importance for Egypt and the Arab world, which face chronic water scarcity, is the explanation of how solar energy can facilitate desalination, the conversion of seawater to fresh water.

The book ends with an ingenious study of how proven solar hot water heating technology can be adapted for large-scale poultry production. By using computer simulations, the authors show that not only is it technically feasible to substitute solar energy for much of the diesel fuel currently used to warm hen houses, but converting to solar makes compelling business sense for poultry farmers as well.

Dr. El-Haggar and his co-authors have deftly bridged the gap between the academic science of solar energy and the practical applications so desperately needed by underserved populations. This valuable handbook can lead to more rapid dissemination of solar technologies in the region and to better health, educational and economic opportunities for the rural populations of Egypt and the Arab world.

Marc Rauch
The American University in Cairo
New York
USA

PREFACE

This book is unique in that, as far as the authors' knowledge, it is the only basic comprehensive guide to solar energy and the potential use of solar powered technologies for sustainable development, primarily on a grass-roots level. The book provides an introduction to basic concepts of solar energy, describes the mechanisms and benefits of related technologies, and suggests potential uses on a practical level by examining case studies. Detailed economic feasibility studies are presented in this book to encourage the implementation of sustainable technologies in the field of solar energy and help communities, especially in the Arab world, live a more sustainable life. The focus on the Arab world in this book stems from the noticeable changes in the environmental and political climates there which provide a great opportunity for the harnessing of renewable energy.

This book strikes a balance between simplicity and comprehensiveness in the fact that although no specialty background knowledge is needed to read it yet it offers broad understanding of solar energy and its applications.

The urgency of exploring alternative energy sources, especially in a region so detrimentally affected by current energy practices on environmental, humanitarian and political levels makes this book a crucial element in raising awareness and activism in the field.

Salah El-Haggar
Department of Mechanical Engineering
The American University of Cairo
Egypt

Sarah Mousa
Arab Studies and Development
Founder of Shamsina, Wynantskill, New York
USA

&

Mohamed El-Morsi
Department of Mechanical Engineering
The American University of Cairo, Ain Shams University
Egypt

ACKNOWLEDGEMENTS

The authors would like to extend our gratitude to everyone at Bentham Science who were involved in the publication of this book. Without their help and dedicated efforts, this book would not have come to fruition. We would also like to thank Humaira Hashmi, Editorial Manager Publication and Asma Ahmed, Manager Publication, for encouraging us to publish this book.

Our sincere appreciation goes to Dr. Khaled Nassar, Eng. Ghada Abdel Azim and Eng. Heidi El-Zanati for their contribution in Chapter 6, Chapter 8 and Chapter 9, respectively.

CONFLICT OF INTEREST

The authors confirm that they have no conflict of interest to declare for this publication.

List of Contributors

Salah El-Haggar	Department of Mechanical Engineering, The American University of Cairo, Cairo, Egypt
Sarah Mousa	Arab Studies and Development, Founder of Shamsina, 5 Fox Glove Court, Wynantskill, 12198, New York, USA
Mohamed El-Morsi	Mechanical Engineering, Power Department, Ain Shams University, Cairo, Egypt Department of Mechanical Engineering, The American University of Cairo, Cairo, Egypt
Khaled Nassar	Department of Construction Engineering, The American University of Cairo, Cairo, Egypt
Ghada Abdel Aziem	The American University, Gleeds Construction Consultancy, Cairo, Egypt
Heidi El Zanaty	Mechanical Engineering, The American University, Technical Trading Company-Tecnco for Poultry Equipment, Cairo, Egypt

ABOUT THE AUTHORS

Dr. Salah M. El-Haggar, is a Professor of Energy and Sustainable Development and Chair of Mechanical Engineering Department, He received a B.Sc. and M.Sc. degree in Mechanical Power Engineering from Ain Shams University and Ph.D. from Washington State University, USA in 1983. He joined The American University in Cairo in 1987.

Dr. Haggar's has more than 30 years' experience in Energy and Environmental Consulting and University teaching. Dr Haggar has been a visiting professor at Washington State University and at University of Idaho, U.S.A. Dr. El-Haggar received the outstanding teaching award from AUC in 1995 as well as a number of outstanding trustees awards. In 2009, Dr. El-Haggar received excellence in Research and creative endeavour award from AUC. In addition Dr. Haggar has 190 Scientific Publications in Environmental, Sustainable Development and Energy Fields, 51 invited presentation, 57 technical report, 16 books, one chapter in 6 different books.

Dr's Haggar Environmental Consulting experience include 5 mega projects for Sustainable Building according to LEED certification (DAR premises, Mall of Egypt MOE, Credit Agricol CA, Union National Bank UNB, Zewail University), more than 40 Environmental/Industrial Auditing for major industrial identities, 20 Compliance Action Plan, 9 Environmental Impact Assessment in addition to his extensive consulting experience in Environmental Engineering, Environmental Auditing, Coastal Zone Management, Environmental impact assessment EIA, Environmental management systems EMS, Energy management, Hazardous and non-Hazardous Waste management, Recycling, Pollution prevention and waste minimization, zero pollution, Biogas/Solar/Wind technology, Community/Desert development, Solid and industrial waste, environmental assessment for the local government and private industries.

Dr. El-Haggar developed a new concept to assess the life cycle of a product according to cradle-to- cradle concept instead of cradle-to-grave concept used worldwide until today. This concept was developed based on 25 years of intensive and continual effort at AUC in the area of waste management with all types of wastes to prove the practicality of this concept from engineering point of view. Recently, a formula and indicator for sustainable development was developed for the first time worldwide according to cradle-to-cradle concept. The formula was included in his recent book published by Elsevier Academic Press entitled "Sustainable Industrial Design and Waste Management: Cradle-to-Cradle for Sustainable Development". This book is the first book worldwide to demonstrate the concept of cradle-to-cradle to protect not only the environment but also the natural resources in terms of business plans.

Sarah Mousa is a 2011 Fulbright Scholar to Egypt, where she studied the potential of solar technologies for use in underprivileged communities. She is the founder of Shamsina, a social enterprise based in Al-Darb Al-Ahmar that produces low-cost solar powered water heaters from local materials.

Sarah obtained Master of Arts in Arab Studies and Development from Georgetown University in 2014; 2011 Fulbright Scholar; 2010 Bachelor of Arts from the Woodrow Wilson School of International and Public Affairs at Princeton University.

Mohamed El-Morsi is an associate professor at the Mechanical-Power Engineering Dept., Faculty of Engineering, Ain Shams University, Cairo, Egypt. Currently, he is a visiting associate professor at the Mechanical Engineering Dept., School of Sciences and Engineering of the American University in Cairo (AUC). He joined AUC in Spring 2008. Dr El-Morsi received his B.Sc. and M.Sc. degrees in Mechanical Engineering from Ain Shams University. In 2002, he received his Ph.D. from the University of Wisconsin-Madison. Later, he spent two and a half years as a postdoctoral student in the Computational Mechanics Center at the University of Wisconsin-Madison. During this time he was working on CFD modeling for immersion lithography. In 2007, he was awarded the Chevening Fellowship from the Foreign & Commonwealth Office, UK to study energy efficiency for three months at the Institute of Energy and Sustainable Development, De Montfort University, Leicester, UK. Since 2007, he has been involved in many activities related to solar and renewable energy. He is also a consultant for the Egyptian Co. for Water Desalination, Taqamisr and Solamisr, Cairo, Egypt. Dr El-Morsi is one of the co-founders of the Solar Energy Development Association. This is a non-profit association developed in coordination with the GIZ PSDP for the promotion of solar energy in Egypt.

INTRODUCTION

The current use of fossil fuels as the main energy source is unsustainable and is associated with both humanitarian and environmental insecurity. According to the International Energy Agency (IEA) the world global energy demand increased from 1990 to 2000 by 14.69% and from 2000 to 2010 by 27.17%. This current rate of usage is alarming as this growing energy demand can outstrip fossil fuel energy supply. This surge in energy demand results in an increase in greenhouse gas emissions that contribute to environmental degradation despite all international efforts to curb the rise of greenhouse gas emissions. Proposals to lower emissions are often met with controversy due to the argument that reducing fossil fuel usage will be associated with decreased development rates. Renewable energy, however, presents a viable alternative to traditional energy sources, especially on a small-scale level. As the world continues to consume fossil fuels and harm its own land and populations the resultant climate change and associated rise in sea level will first and foremost affect vulnerable countries like those on the Mediterranean. The expanding world population, with its increasing development, especially in the Middle East and North Africa (MENA) is using fossil fuels at an untenable rate. Despite the vast potential for renewable energy which the MENA region possesses, exploitation, conflict and instability associated with energy resources remain a major political and economic obstacle to development. The MENA region lies on the global sun-belt and is potentially one of the most productive regions for solar energy usage. The possibility for solar technologies to improve living conditions by providing a clean, sustainable energy source and expand local employment opportunities is especially significant among remote and low-income communities in this region. There is potential for great enthusiasm for alternative energy sources throughout the MENA region, as displayed by select underprivileged communities which have experimented and embraced solar technologies.

The objective of this book is to provide an overview of solar energy, present several solar technologies and suggest applications that can greatly benefit low-income, developing communities in the geophysical, societal and cultural context of MENA.

The book begins by providing a brief introduction about solar radiation and climate change in chapters one and two, respectively. Chapter three then presents the negative impacts of fossil fuels and how can solar energy offer a viable alternative. Subsequently chapter three provides a brief overview of selected solar powered technologies, including their mechanisms and possible applications. The implementation of these technologies is presented in Chapter five. Chapter six then focuses on natural lighting and studies its importance, and significant influence on electric bills.

The energy solutions in MENA which is a leading candidate for solar energy use are discussed in Chapter seven. Finally, the last two chapters in the book present economic feasibility studies that are necessary to encourage the implementation of sustainable technologies in the field of solar energy in order to achieve more sustainable communities in the MENA region.

SUSTAINABLE SOLAR ENERGY SYSTEMS

Challenges and Economics for the Arab World

CHAPTER 1

Sunlight and Solar Radiation

Sarah Mousa[2,#], **Mohamed El-Morsi**[1] and **Salah El-Haggar**[1,*]

[1] *The American University in Cairo, Department of Mechanical Engineering, Cairo, Egypt*

[2] *Arab Studies and Development, 5 Fox Glove Court, Wynantskill 12198, New York, USA*

Abstract: The sun is a hot sphere of gas with an internal temperature of 15 million degrees Celsius, the surface of the sun is at a temperature that is approximately 5500 Kelvin. The total power emitted by the sun is 3.83×10^{26} W, and the intensity is about 6.33×107 W/m^2. The solar radiation reaching the earth's surface, 1367 W/m^2, is reduced by absorption, reflection and scattering of the atmosphere. Solar radiation and its intensity on earth's surface varies based on latitude and longitude location, time of year, time of day, local atmospheric variations such as vapor, cloud or pollution concentration, and atmospheric effects such as absorption and scattering. All these are crucial factors when considering the optimum design and location for solar powered technologies. This chapter will help us consider where, when and how solar radiation should be harnessed in order to take the greatest advantage of energy reaching the earth's surface.

Keywords: Air Mass, Atmospheric gases, Electromagnetic spectrum, Extraterrestrial, Gamma rays, Infrared, Intensity of solar radiation, Mie scattering, Photon, Photosphere, Power emitted by the sun, Radio waves, Rayleigh scattering, Solar insolation, Sun internal temperature, Sun surface temperature, Ultraviolet, Visible light, X-rays.

INTRODUCTION

This chapter provides sunlight and solar radiation background information needed

* **Corresponding author Salah El-Haggar:** Department of Mechanical Engineering, School of Sciences and Engineering, The American University in Cairo, AUC Avenue P.O. Box 74, New Cairo 11835, Egypt; Email: elhaggar@aucegypt.edu.

Founder of Shamsina

Salah El-Haggar, Sarah Mousa & Mohamed El-Morsi (Eds.)
All rights reserved-© 2016 Bentham Science Publishers

for understanding the basic mechanisms of solar powered technologies and how they can be used for sustainable development. This includes the features of the sun, the light and energy which it provides, and the factors which impact the intensity of this energy on earth's surface.

SUNLIGIIT

The sun is a hot sphere of gas with an internal temperature of 15 million degrees Celsius [1]. Nuclear fusion reactions at the core which convert hydrogen into helium contribute to the high internal temperatures. The surface of the sun is called the photosphere and its temperature is approximately 5500 Kelvin [2]. Sunlight is a form of electromagnetic radiation (an energy form that travels in waves through space) and can be described in terms of an electromagnetic spectrum with light consisting of certain wavelengths. The power emitted from the sun is composed of a wide range of wavelengths in the electromagnetic spectrum, and appears white or yellow to the human eye [3].

Visible light is only one subset of a spectrum which constitutes light emitted by the sun. Fig. (**1.1**) illustrates the electromagnetic spectrum and indicates wavelengths which correlate with each segment of the spectrum. It is notable, as indicated in Fig. (**1.1**), that shorter wavelengths are correlated with lower energy.

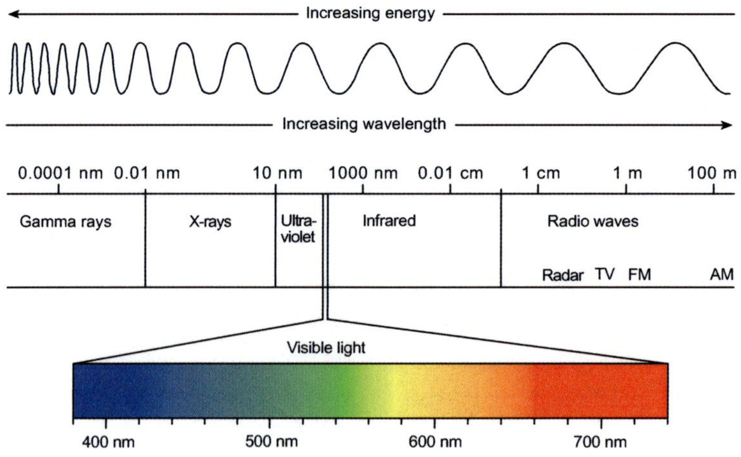

Fig. (1.1). The electromagnetic spectrum [4].

In addition to wavelengths, sunlight can also be described in the form of packets of energy called photons. A photon is characterized either by its wavelength (λ) or energy (E). Wavelength and energy have an inverse relationship, as shown by Eq. (**1-1**) [3].

$$E = \frac{hC}{\lambda} \tag{1-1}$$

In Eq. (1-1), (h) represents Plank's constant and (C) is the speed of light. According to this formula, the longer the wavelength of a photon, the lower the corresponding energy is, and the shorter a wavelength is, the higher the corresponding energy. The unit of energy commonly used to describe photons is the electron-volt (eV), which is the amount of energy needed to raise an electron 1 volt.

The photon flux is relevant when considering the intensity of energy emitted from the sun. The photon flux is the number of photons per unit area per second; the concept will be important in photovoltaics when considering the electrons generated and thus current produced, by a solar cell. The photon flux, however, does not provide details on photon wavelength or energy, either one of which would additionally be needed to calculate the power density for photons at a particular wavelength. Multiplying the photon flux by the energy of the photons yields the power density. One noteworthy implication is that higher energy photons will require a lower photon flux than lower energy photons to yield a similar power density. Spectral Irradiance (F) is the power density of a photon at a particular wavelength and is another important concept in photovoltaics [3].

The total power emitted by the sun, 3.83×10^{26} W [1], is calculated by multiplying emitted power density by the surface area of the sun, 6 million km^2 [2], and the intensity of the solar radiation, H_{sun}, is about 6.33×10^7 W/m^2, as shown in Fig. (**1.2**). The solar radiation at the top of the earth's atmosphere, extraterrestrial radiation, can be calculated by using the total power emitted by the sun's surface and accounting for the earth's distance from the sun. A yearly average value for the extraterrestrial radiation, H_0, is the solar constant 1367 W/m^2 as shown in

Fig. (**1.2**). The actual power density varies slightly throughout the year since the earth moves in an elliptical orbit, but the seasonal differences are relatively insignificant. Power variation due to the earth's movement in an elliptical orbit is 3.4% [3]. Also, the actual power density varies since the output of the sun changes due to sun spot cycles.

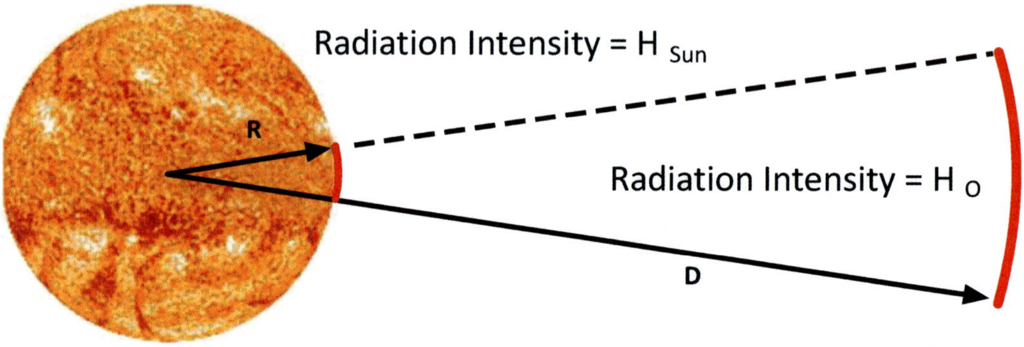

Fig. (1.2). Distance from the sun and solar radiation intensity [5].

SOLAR RADIATION ON EARTH'S SURFACE

The atmosphere has an important impact on the intensity and spectral content of solar radiation reaching the earth's surface. Through absorption, reflection and scattering the atmosphere reduces the solar radiation reaching the earth's surface. Due to greater absorption or scattering of certain wavelengths, the atmosphere also plays a role in determining the spectral content of the solar radiation reaching the earth's surface. The atmosphere of the earth consists of several spheres—the troposphere, stratosphere (which includes the ozone layer), mesosphere, thermosphere, exosphere and magnetosphere [6].

Atmospheric gases including Ozone (O_3), Carbon Dioxide (CO_2) and Water Vapor (H_2O) are particularly important in that they highly absorb photons that have energies close to their bond energies. Water Vapor and Carbon Dioxide, for example, absorb infrared light, IR, while Ozone absorbs ultraviolet light, UV, [7]. Although the absorption of the UV light is not complete, hence harmful effects of ultraviolet light on earth's surface such as sunburn and skin cancer.

The major factor reducing solar radiation on the earth's surface, however, is absorption and scattering of light caused by air molecules and dust. This process is evident as the path length through the atmosphere changes throughout the day. For example, at midday, when the path length of sunlight to the earth's surface is most direct the incident light appears white. For longer paths, during sunrise and sunset, higher energy photons are more effectively absorbed and thus the redder appearance of the sun earlier and later during the day [3].

One mechanism for the scattering of light as it moves through the atmosphere is called Rayleigh scattering and is caused by small molecules in the atmosphere. This has a larger effect on high energy (short wavelength) photons because of the similarity in size to the atmospheric molecules. A second type of scattering, Mie scattering, is caused by aerosols and dust particles [7].

The solar radiation at different geographical locations on earth's surface varies greatly due to several factors. These variations are based on latitude and longitude location, time of year, time of day, local atmospheric variations such as vapor, cloud or pollution concentration, and atmospheric effects such as absorption and scattering. Regions close to the equator tend to have low variability due to seasons because of the relative constancy in atmosphere. Desert regions, which are close to the equator and have lower concentrations of clouds, have both higher amounts of and lower variability in solar radiation throughout the year. Areas with high cloud cover receive significantly lower concentrations of solar radiation. The map featured below in Fig. (**1.3**) indicates the amount of solar radiation reaching the continental earth. As indicated, an area such the Sahara desert is among those with the highest concentrations of average solar radiation.

The time of day significantly impacts the amount of solar radiation reaching a particular point on earth. When considering the optimal efficiency of solar powered technologies this concept is important to take into account. Air Mass, (AM) the shortest path of solar radiation to a particular point on earth, provides a quantitative approach to accounting for the time of day. Air Mass is defined by Eq. (1-2), where (θ) is the angle of the sun from the vertical. According to this formula, Air Mass is equal to 1 when the sun is directly overhead and thus when solar radiation is highest. As the sun moves across the sky, solar radiation is lowered [7].

$$AM = \frac{1}{cos\theta} \quad \textbf{(1-2)}$$

The motion of the sun (or rather, the elliptical orbit of the earth) has a significant impact on the amount of energy that will reach a solar collector [3]. It is also of interest to note, that while the solar radiation reaching the earth may vary, the ground itself maintains a constant temperature throughout the year. While the surface ground undergoes solar radiation absorption and losses to the ambient air, deeper ground absorbs and maintains constant solar radiation due to the fact that upper layers of ground serve as insulation [3]. This notion will be important in considering how solar energy stored in the ground can be exploited, particularly in solar building designs.

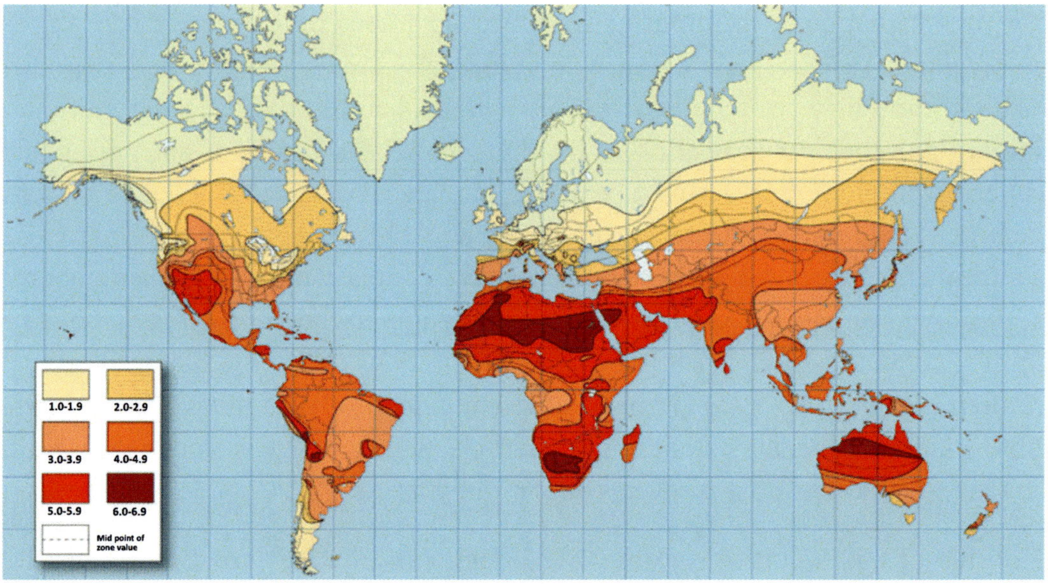

Fig. (1.3). Solar insolation (kWh/m^2/year) on an optimally tilted surface during the worst month of the year [8].

DISCUSSION

Solar radiation and its intensity on earth's surface with varying geographical location, season and time of day, as examined in this chapter, are crucial factors when considering the optimum design and location for solar powered technologies and identifying challenges that must be overcome. This chapter will help us consider where, when and how solar radiation should be harnessed in order to take the greatest advantage of energy reaching the earth's surface. The factors discussed here will serve as a base for the rest of the book, and especially Chapter 4 on basic designs of solar powered technologies. Section 0, on the role of the atmosphere in absorbing solar radiation will be important to remember in Chapter 2 on Climate Change, which will highlight the essentiality of the atmosphere for healthy life on earth.

QUESTIONS

1. What does the electromagnetic spectrum represent?
2. What determines the power density of solar radiation at a specific location?
3. How does the atmosphere impact solar radiation on earth's surface?
4. What are the determining factors of intensity of solar energy at a specific geographic location?
5. How do season and time of day alter solar energy intensity?
6. Which geographic locations enjoy the highest intensity of solar energy?

REFERENCES

[1] Our Solar System, "US National Aeronautics and Space Administration (NASA) Central Operation of Resources for Educators (CORE)", *Available: http://solarsystem.nasa.gov/multimedia/downloads/ SolarSystem_FC.pdf.*Accessed: 27-Jun-2015 [Online]

[2] Solar System Exploration: Planets: Sun: Facts & Figures, "US National Aeronautics and Space Administration (NASA)", *Available: http://solarsystem.nasa.gov/planets/profile.cfm?Display =Facts&Object=Sun.*Accessed: 27-Jun-2015 [Online]

[3] C. Honsberg, and S. Bowden, PVCDROM, "PV Education", *Available: http://www.pveducation. org/pvcdrom.*Accessed: 27-Jun-2015 [Online]

[4] The Electromagnetic Spectrum, "Cyberphysics", *Available: http://www.cyberphysics.co.uk/topics/ radioact/Radio/EMSpectrumcolor.jpg.*Accessed: 27-Jun-2015 [Online]

[5] S. Niazi, *Earth Changes..* Word Press, 2012.

[6] M. Beychok, Earth's Atmosphere, "The Encyclopedia of Earth (EoE)", *Available: http://www.eoearth.org/view/ article/171011/,* 2011.Accessed: 27-Jun-2015 [Online]

[7] J.A. Duffie, and W.A. Beckman, *Solar Engineering of Thermal Processes..* 4th ed John Wiley & Sons: Hoboken, NJ, USA, 2013. [http://dx.doi.org/10.1002/9781118671603]

[8] The Ugly Side of Solar Panels, "LOW-TECH MAGAZINE", *Available: http://www.lowtechmagazine. com/ 2008/03/the-ugly-side-o.html.*Accessed: 18-Jun-2015 [Online]

CHAPTER 2

Climate Change

Sarah Mousa[2,#], **Mohamed El-Morsi**[1] and **Salah El-Haggar**[1,*]

[1] *The American University in Cairo, Department of Mechanical Engineering, Cairo, Egypt*

[2] *Arab Studies and Development, 5 Fox Glove Court, Wynantskill 12198, New York, USA*

Abstract: The earth's temperature is determined by the incoming radiation from the sun and the outgoing infrared radiation emitted by the earth. Radiation emitted by the earth is largely dependent on the composition of the earth's atmosphere. The accumulation of greenhouse gases, due to human activities, in the earth's atmosphere absorbs infrared radiation emitted by the earth's surface and keeps it in the atmosphere. With today's rate of fossil fuel utility, compounds are released into the atmosphere, soil and seas on a daily basis, resulting in significant changes in the atmosphere. In recognition of how damage caused by fossil fuels harms an environment shared by all, the United Nations Framework Convention on Climate Change (UNFCC) put forth the Kyoto Protocol in 1997. Climate change is among the many reasons, which make increased research on and immediate implementation of solar powered technologies not simply a luxury, but a necessity for the future environmental wellbeing of earth. This chapter presents the problem of global warming, its reasons and efforts led by decision makers to seek various approaches to tackle the core of the problem and mitigate its severe environmental, economic and social impacts.

Keywords: Anthropogenic gases, Cap-and-trade, Carbon credit, Carbon dioxide, Certified emission reductions, Clean development mechanism, European Union allowances, Global warming, Greenhouse gases, Kyoto Protocol, Terrestrial temperature.

* **Corresponding author Salah El-Haggar:** Department of Mechanical Engineering, School of Sciences and Engineering, The American University in Cairo, AUC Avenue P.O. Box 74, New Cairo 11835, Egypt; Email: elhaggar@aucegypt.edu.

Founder of Shamsina

Salah El-Haggar, Sarah Mousa & Mohamed El-Morsi (Eds.)

All rights reserved-© 2016 Bentham Science Publishers

INTRODUCTION

It is important to understand the state of the existing climate and the effect of the current energy usage on it before discussing solar powered technologies and the benefits associated with their use. Fossil fuels, including coal, petroleum and natural gas, are currently the most prevalent source of energy worldwide. These sources contribute to global warming as will be discussed in this chapter. This chapter will also address the international responses to this phenomenon.

GLOBAL WARMING

The temperature on the earth's surface is closely associated with weather patterns which make it inhabitable for living beings. This temperature is determined by two aspects: first, the incoming radiation from the sun and second, the outgoing infrared radiation (heat) emitted by the earth. Radiation emitted by the earth is largely dependent on the composition of the earth's atmosphere. If there were no atmosphere, earth's temperature would be -18°C whereas the earth's average temperature is 15°C [1].

The presence of CO_2 in the earth's atmosphere plays a crucial role in decreasing radiation emitted from earth. Fossil fuels contain a high amount of carbon. Human activities involving fossil fuel energy result in the release of anthropogenic gases (including CO_2, methane, ozone, nitrous oxides and chlorofluorocarbons) which impact the earth's atmosphere. The resulting increasing amount of gases in the atmosphere absorbs infrared radiation emitted by the earth's surface and keeps it in the atmosphere instead of permitting it to be radiated to deep space. The release of these pollutants, or greenhouse gases, through the use of fossil fuels is a routine occurrence, and can be either expected or accidental. With today's rate of fossil fuel utility, compounds are released into the atmosphere, soil and seas on a daily basis, resulting in significant changes in the atmosphere [2]. The greenhouse gas (GHG) emissions has been increasing almost exponentially. According to the World Bank indicators the GHG emissions has increased from 23 Gt in 1990 to 35 Gt in 2011, which is about 53% increase.

Other human activities which contribute to the increase of CO_2 levels in the atmosphere include the destruction of forests, since this reduces subsequent rates of photosynthetic CO_2 assimilation until ground cover is restored. The change in atmosphere caused by these factors potentially increases terrestrial temperatures, threatening the environmental status quo. Fig. (**2.1**) indicates the upward shift in CO_2 concentrations over the past century, and particularly in recent decades, and the associated rise in terrestrial temperatures [1].

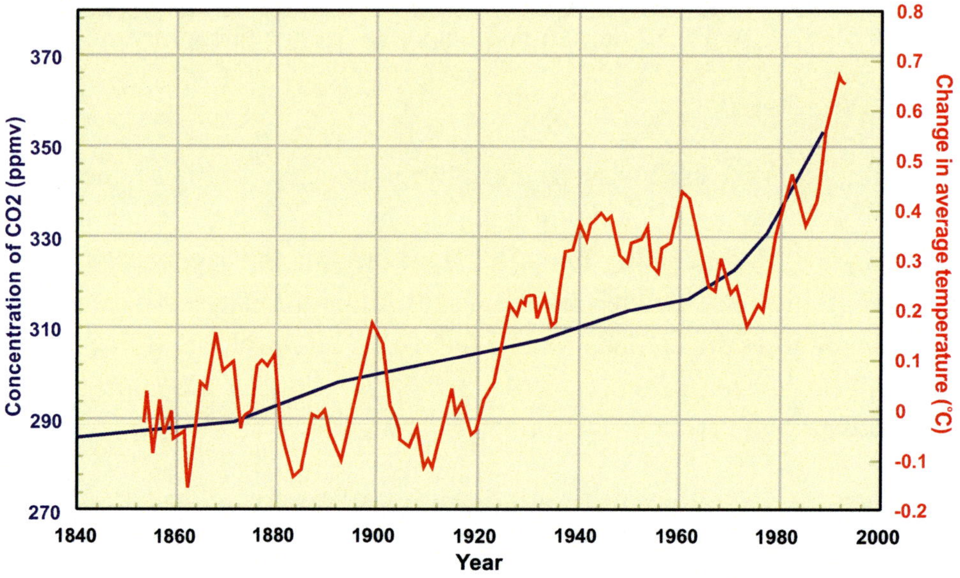

Fig. (2.1). Concentration of CO_2 and average terrestrial temperature [1].

Some studies suggest that by 2030, CO_2 levels could double—leading to an upward shift in temperature ranging from 1-4°C. This in turn could affect wind patterns and rainfall, causing a rise in sea levels and the interior of continents to dry [1]. The International Energy Agency (IEA) announced that carbon-emissions reached a record high in 2011 [3]. Along with a continued rise in temperature, rise in sea water levels, intensified storms, crop failures and increased droughts are among the expected repercussions of climate change and are already evident [4].

Opponents of the notion of global warming suggest that if CO_2 levels and the temperature of the earth are considered over thousands of years, a natural cycle of rising and falling temperatures exists. While there is indeed a clear pattern of rising and falling temperatures, the fact remains that in recent decades CO_2 levels have risen dramatically and that there seems to be a correlation between CO_2 emissions and temperature. While there are extreme arguments on both sides of the issue, global warming is a pressing man-made issue for which we must explore solutions.

KYOTO PROTOCOL

In recognition of how damage caused by fossil fuels harms an environment shared by all, the United Nations aims to fight global warming using several methods. The United Nations Framework Convention on Climate Change (UNFCC) put forth the Kyoto Protocol in 1997. It is an international environmental treaty the goal of which is the "*stabilization of greenhouse gas concentrations in the atmosphere at a level that would prevent dangerous anthropogenic interference with the climate system.*" The Protocol is adopted in Kyoto, Japan and came into force in 2005. To date, 191 countries have signed and ratified the treaty [5].

The Protocol is derived from the notion that human actions are responsible for an increase in the temperature of the earth, as discussed in the previous section, and that this will have a detrimental effect on the livelihood of living beings. Kyoto's objective is to present a legally binding agreement through which signatories must minimize greenhouse gas emissions. The protocol aimed to reduce greenhouse gases by 5.2% from 1990 to 2012.

The Protocol was viewed as controversial by some groups, particularly in the United States which has not signed or ratified the protocol. While Kyoto does imply long-term benefits, the short-term costs are seen as too drastic by opponents. A decrease in development rates, job losses and other associated costs are among the consequences raised be critics. Opponents also point to developing countries and the increased difficulty that these nations, which often lack the financial or structural capability to obtain more efficient technologies or explore environmentally-friendly alternatives, will face in working to decrease greenhouse gas emissions [6].

CARBON CREDITS

One of the outcomes of Kyoto protocol is the initiation of carbon credit markets, where carbon has an "economic value" and can be traded or banked by governments, firms or other bodies. In addition, a new development mechanism called the Clean Development Mechanism (CDM) has been introduced to allow the industrialized countries included in the protocol to invest in cleaner projects that reduce emissions in developing countries. This could be less costly than implementing emission reduction mechanisms in their own countries. This mechanism creates a win-win situation: the developing countries in which such projects are established would benefit from the cleaner technology and create jobs, while the industrialized countries would receive carbon credits for meeting their emission reduction targets [7].

The concise definition of carbon credits may certainly vary from one reference to another; while basically all the reference would highlight one basic concept in common, which is the relevance of carbon to an economic value. Firstly, the Credo Reference provides the definition of carbon credits as "qualified and verified reductions in greenhouse gas emissions, that are tradable and have financial value and (in most cases) created under a legal framework for greenhouse gas trading, such as the Kyoto Protocol, or the EU Emissions Trading Scheme" [8]. Sometimes, these emission reductions could be done "voluntarily" by any organization and without any legal commitment. Given the highly negative impacts of the global warming on the environment, a few legislative systems have been established, such as the EU Greenhouse Gas emission trading scheme and the Kyoto Protocol, which assist sustaining the carbon credits market. Carbon credits are provided in the form of certificates, and are measured in terms of Certified Emission Reductions (CER's), each one of which corresponds to the reduction of one ton of carbon dioxide emission [8].

The CER's is the acronym linked to the Clean Development Mechanism projects under the Kyoto Protocol [8]. Other acronyms are: European Union Allowances (EUA's), which are linked to the EU greenhouse gas emission trading scheme, Emission Reduction Units (ERU's) linked to the Joint Implementation (JI) projects, under the Kyoto Protocol, Assigned Allowance Units (AAU's) associated with the International Emissions Trading under the Kyoto Protocol, and finally, Verified Emissions Reductions (VER's) which are certified and given to the voluntary initiatives. The greenhouse gases are not solely about carbon chemical compounds, however, the carbon term refers to the carbon element in the carbon dioxide formula, which is the most abundant greenhouse gas in the global atmosphere [8].

In the United States the trading of carbon falls under the scheme known as the "cap-and-trade" system, where the carbon dioxide emissions are to be "capped" [9]. Briefly, this approach is about a certain limit for each "producer"; hence, those who do not exceed their own limit can trade their remaining allowance of production to those, who exceeded their limit and still need the extra allowance [9]. According to the National Allocation Plan (NAP), each market participant is "allocated" a certain limit of greenhouse gases that cannot be exceeded, and that is the "so-called-cap" [9]. When the participants exceed their limits and make deals with those, who still have remaining "allocation", that is the "so-called-trade" [9].

DISCUSSION

The currently increasing use of fossil fuels as a main energy source is unsustainable. Not only are fossil fuels nearing depletion, but also potential environmental devastation adds urgency to the search for alternative options. Climate change is among the many reasons, many more of which will be discussed in Chapter 3, which make increased research on and immediate implementation of solar powered technologies not simply a luxury, but a necessity for the future environmental wellbeing of earth.

QUESTIONS

1. What determines temperature on the earth's surface?
2. How does Carbon Dioxide impact solar radiation on earth's surface? How does it impact terrestrial temperature?
3. What human activities have an effect on the amount of Carbon Dioxide on earth's surface?
4. What are the dangers of global warming to the environment? To humanity?
5. How has the international community addressed global warming? To what extent have international efforts been successful?
6. How has Egypt addressed climate change?

REFERENCES

[1] C. Honsberg, and S. Bowden, PVCDROM, "PV Education", *Available: http://www.pveducation.org/pvcdrom.*Accessed: 27-Jun-2015 [Online]

[2] *World Energy Assessment Energy and the Challenge of Sustainability*. Bureau for Development Policy, United Nations Development Programme, 2000.

[3] Global carbon-Dioxide Emissions Increase by 1.0 Gt in 2011 to Record High, "The International Energy Agency (IEA)", *Available: https://www.iea.org/newsroomandevents/news/2012/may/global-carbon-dioxide-emissions-increase-by-10-gt-in-2011-to-record-high.html.*Accessed: 27-Jun-2015 [Online]

[4] Meeting the Climate Change Challenge Sustaining Livelihoods, "Global Environment Facility's (GEF)", *United Nations Development Programme (UNDP),* 2004.

[5] Kyoto Protocol, "The United Nations Framework Convention on Climate Change (UNFCC)", *Available: http://unfccc.int/essential_background/kyoto_protocol/items/1678.php.*Accessed: 27-Ju--2015 [Online]

[6] E. Bond, "Climate Change and the Kyoto Protocol", [Online]. Available: http://climatechange.sea.ca/kyoto_protocol.html. [Accessed: 27-Jun-2015].

[7] A. Mishra, R. Jain, H. Afrin, and A. A. Sinha, "Carbon Credit for Sustainable Developement", *Recent Res. Sci. Technol.,* vol. 6, no. 1, 2014.

[8] S. Habbitts, "Carbon Credits", In: *The A to Z of Corporate Social Responsibility: A Complete Reference Guide to Concepts, Codes and Organisations..* John Wiley & Sons, 2009.

[9] G.M. Padis, "Carbon Credits as Collateral", *J. Tech. L. Pol'y,* vol. 16, p. 343, 2011.

CHAPTER 3

Solar Energy and Associated Benefits

Sarah Mousa[2,#], **Mohamed El-Morsi**[1] and **Salah El-Haggar**[1,*]

[1] *The American University in Cairo, Department of Mechanical Engineering, Cairo, Egypt*

[2] *Arab Studies and Development, 5 Fox Glove Court, Wynantskill 12198, New York, USA*

Abstract: The environmental problems associated with the use of fossil fuels can be largely diminished by changing to renewable energy sources. Traditional energy sources are harmful to the environment on both local and global scales, largely because their use is associated with the emission of greenhouse gases. Solar power systems, along with those which rely on other renewable energy sources such as wind and hydroelectric power, produce minimal carbon emissions. Considering its abundance, solar energy presents itself as a clean viable main energy source alternative to traditional reliance on fossil fuels. On the economic level, solar energy has the potential to serve as a stimulant for economic growth and sustainable development. Throughout the world, and particularly in developing countries, access to energy is highly correlated with social and economic factors. Access to affordable energy for all is crucial for economic growth. Also, the social benefits associated with the usage of solar energy are potentially most significant for underserved populations in developing countries. This chapter shows how solar energy presents itself as an alternative to traditional energy sources that is healthier for both the environment and humanity, and has the potential to enhance the social and economic fabric of a community.

Keywords: Air pollution, Global mean temperature, Health, Human development index, Respiratory diseases, Solid fuel.

INTRODUCTION

This chapter continues the discussion started in the previous chapter, Chapter 2,

* **Corresponding author Salah El-Haggar:** Department of Mechanical Engineering, School of Sciences and Engineering, The American University in Cairo, AUC Avenue P.O. Box 74, New Cairo 11835, Egypt; Email: elhaggar@aucegypt.edu.

Founder of Shamsina

Salah El-Haggar, Sarah Mousa & Mohamed El-Morsi (Eds.)
All rights reserved-© 2016 Bentham Science Publishers

about the negative impacts of fossil fuels and presents the benefits associated with using solar energy and related technologies. It is interesting to know that the amount of solar energy that reaches earth every hour is greater than the energy that the world's population consumes in a year [1]. The benefits associated with the exploitation of this energy, both to the environment and humanity, are innumerable. More details and examples about this clean source of energy are discussed in this chapter.

SOLAR POWER

Solar power is the conversion of sunlight into electricity. This can be accomplished by several different means, one of which is Concentrated Solar Power (CSP), a direct solar thermal process. Several solar powered technologies rely on CSP, including solar cookers, dryers, and water heaters. Solar power can also be directly converted into electricity, as is the case with photovoltaic panels. Solar cells which comprise these panels directly convert sunlight into electricity. Considering its abundance, solar energy presents itself as a viable alternative to traditional reliance on fossil fuels as a main energy source [2].

Fossil fuels, including coal, petroleum and natural gas, are problematic on several fronts. Fossil fuels are associated with a number of hazards relating to the environment, human health, security and economic stability. These hazards place increased pressure on a need for investment in renewable energy sources. The remainder of this chapter will discuss the benefits of solar power, especially as opposed to traditional energy sources. It will present a case for increased use of solar power technologies, not as an end goal in and of itself, but as a means for protecting the environment and public health and as a catalyst for social and economic development.

ENERGY AND ENVIRONMENT

The environmental problems associated with climate change, discussed in Chapter 2, can be largely diminished by transitioning from fossil fuels to renewable energy sources. Traditional energy sources are harmful to the environment on both local and global scales, largely because their use is associated with the emission of greenhouse gases. Environmental degradation is felt even more strongly in developing countries, which struggle to respond to climate change and rely on environmentally sensitive activities that are especially harmed by fossil fuel use, such as agriculture and fishing.

Solar power systems, along with those which rely on other renewable energy sources such as wind and hydroelectric power, produce minimal carbon emissions. Solar power is thus a clean energy source, associated with no harm to the environment. Wiser use of this natural resource can result in a lower dependence on fossil fuels, thus reducing emissions and improving both the local and global environment.

Fig. (**3.1**), shows predicted climate change over a century for given various scenarios. In a world that seems headed for rapid development (A1), this graph makes clear the pressing need for increased use of alternative energy technologies (scenario B1). Special consideration for the amount of harm that a change in temperature, by even a single degree, could result in making the graph below a stark warning for continued fossil fuel-reliant global development.

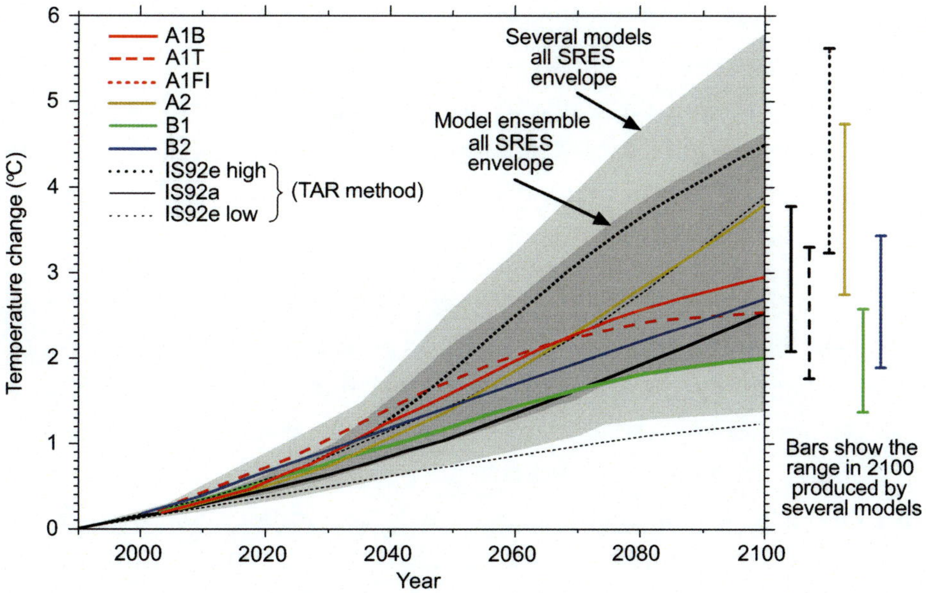

Fig. (3.1). Global mean temperature change and future changes for six illustrative special report on emissions scenarios (SRES) [3].

A1: A world of rapid economic growth and rapid introductions of new and more efficient technologies, A1B: (balanced), A1FI: (fossil intensive), A1T: (non-fossil).

A2: A very heterogeneous world with an emphasis on family values and local traditions.

B1: A world of "dematerialization" and introduction of clean technologies.

B2: A world with an emphasis on local solutions to economic and environmental sustainability.

IS92a: "business as usual" scenario.

ENERGY AND HEALTH

The adverse health effects caused by use of traditional energy sources are tangible in every facet of a community. Air pollution impacts urban as well as rural communities and is felt both outdoors and indoors. Although often overlooked, fossil fuels are associated with a number of harmful health effects within households. Women and young children suffer the highest exposure from household pollution. The main health effects associated with air pollution are: infectious respiratory diseases (including tuberculosis), chronic respiratory diseases (including bronchitis and lung cancer), pregnancy complications, asthma, blindness and heart disease [4].

The adverse health effects of energy processes are felt more heavily in developing countries, where biomass (wood, coal) is more commonly used in poor households. A study conducted in India in 1995 indicated that household use of solid fuels (wood, charcoal, *etc.*) for cooking or heating caused the premature death of 500,000 women and children under the age of 5 per year, comprising approximately 5% of the health burden for that country. The World Health Organization (WHO) estimated that 2.5 million women and children worldwide die prematurely every year due to household use of solid fuels [5].

Other health-damaging effects are associated with solid fuel usage such as wood-burning, especially if that wood is treated with insecticides or lead paint, which are hazardous when inhaled [5]. Coal, a primary energy source until replaced with oil in the 1960s, currently comprises about one-fifth of total worldwide energy consumption. When burned, coal releases carbon monoxide, carbon dioxide and methane which when inhaled can cause drowsiness, headache or suffocation [5].

At the community level, urban air pollution is chief in terms of health-damaging effects of energy usage. The bulk of this pollution is caused by fuel combustion, with power plants and automobiles as the two largest urban polluters. Fuel combustion produces several health-damaging pollutants, mostly small particles. Small particles consist of a mix of primary combustion particles (carbonaceous materials) and secondary conversion products including sulfate and nitrate aerosols. Small particles settle deep into the lungs where their clearance is slow and they can potentially cause damage. In developed countries, a number of measures have been taken in recent years to reduce urban air pollution, including improved fuel efficiency for automobiles [5].

In developing countries, air pollution is generally higher for a number of reasons. In urban centers, automobiles are not as fuel-efficient as they tend to be in developed countries. Other activities which contribute to air pollution include the highly damaging activity of burning garbage. In Cairo, this is done within the city limits and has especially increased in recent years due to the government's mass culling of pigs which formerly consumed organic waste. In rural areas, the seasonal burning of agricultural waste contributes tremendously to air pollution and is increasingly harmful if the land was treated with pesticides [5].

Harmful effects associated with solar energy systems, once in place, are virtually nonexistent. By replacing fossil fuels with solar energy, many of the harmful effects associated with this traditional energy source will be diminished. In the case of automobile exhaust, as it is noted as one of the chief pollutants, cars which run on batteries powered by solar electricity are among potential alternatives.

ENERGY AND SOCIETY

The social benefits associated with the usage of solar energy are potentially most significant for underserved populations in developing countries. There are approximately 2 billion people worldwide who depend on wood, agricultural residue, or dung as their primary sources for cooking and heating. The burden of preparing energy sources falls most heavily on women, who are often responsible for the collection of these materials and are most exposed to the adverse health effects associated with usage of these fuels. Additionally, young children are disproportionately exposed to the negative health effects of these fuels and are at times forced to forgo their education in order to contribute to the family's basic survival. Women and children are consequently denied time for education and income-earning activities.

Access to cleaner, sustainable energy sources, including solar power, could thus have a revolutionizing impact on the lives of all, but especially women and young children in underserved communities. Solar energy systems could provide a dramatic improvement in living conditions, and allow time for members of a community to engage in other activities.

ENERGY AND THE ECONOMY

Solar energy has the potential to serve as a stimulant for economic growth and sustainable development. Throughout the world, and particularly in developing countries, access to energy is highly correlated with social and economic factors. Access to affordable energy for all is crucial for economic growth. For the 2 billion people worldwide living under the poverty line, improved living conditions start with the satisfaction of basic needs such as access to food, running water, housing, health services, education and jobs. Energy plays in important role in providing all of these basic needs [6].

Energy services such as heating, cooking, refrigeration, lighting, and transportation systems are closely tied to the guarantee of basic living conditions. Low energy consumption is not the cause of poverty, but it is associated with many of its elements, including poor education and healthcare [6]. Energy can become a tool in the eradication of poverty when it is targeted specifically towards impoverished, and especially rural, communities.

The Human Development Index (HDI) measured by the United Nations Development Program (UNDP) is an indicator of quality of life, and accounts for average life expectancy, education level and per capita gross domestic product.

Past HDI figures show a correlation with per capita commercial energy consumption. Energy consumption has a strong impact on HDI, especially when energy consumption is below 1,000 KOE (kilogram of oil equivalent) per capita. Fig. (**3.2**) illustrate this relationship [6].

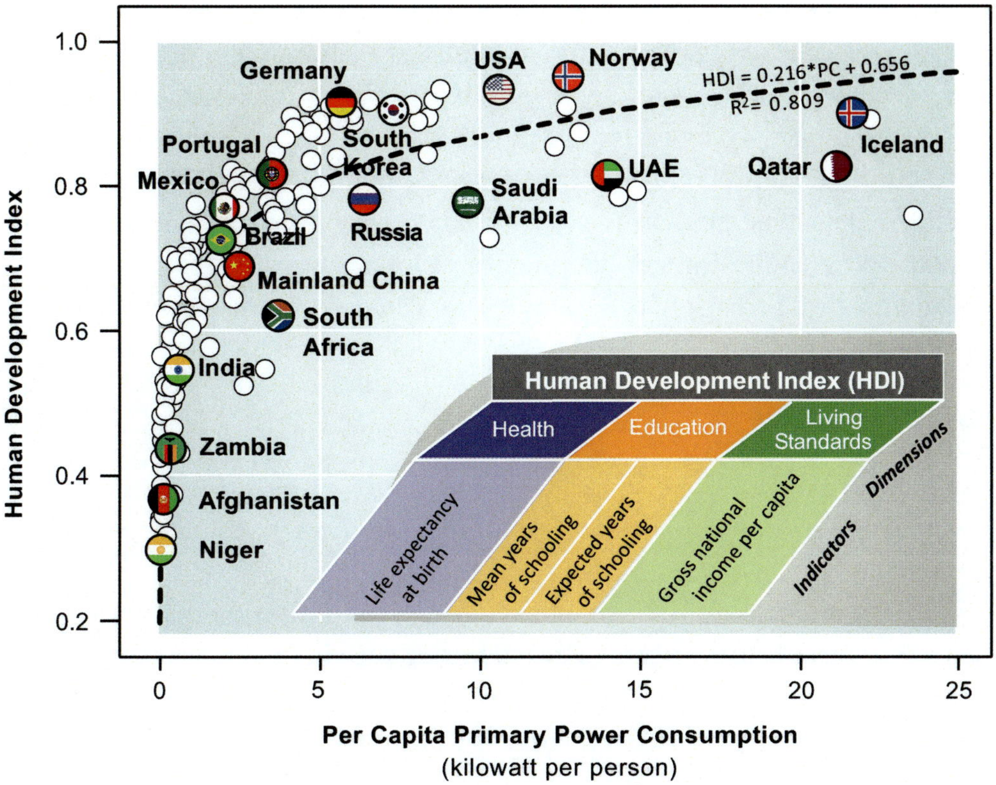

Fig. (3.2). Estimated relationship between human development index (HDI) and per capita energy consumption [7].

The poor, especially in rural areas, pay a very high price for energy—a lack of access to reliable energy sources usually means that people are left to depend on their own labor to meet energy needs. This process is economically costly in terms of human time and labor, and as mentioned in previous sections has additional social and health costs [6]. Environmentally-friendly renewable energy systems can play a role in economic growth not only as an end (*i.e.* in its role in providing needed energy services) but also as a means.

DISCUSSION

In this chapter it has been shown how solar energy presents itself as an alternative to traditional energy sources that is healthier for both the environment and humanity, and has the potential to enhance the social and economic fabric of a community. This is especially the case for those underprivileged communities that do not have reliable access to energy, and thus consume much of their time and efforts in securing sources that are especially detrimental when used. As will be discussed in Chapter 4, several solar-powered technologies are simple, can be locally constructed and would be ideal for such communities. This chapter has also made it clear that the harms of traditional energy sources and potential of solar energy is not limited to underdeveloped societies, and thus the implementation of proposed technologies in Chapter 4 and derivates thereof would benefit virtually all.

QUESTIONS

1. What is Solar Power?
2. How will the continued use of fossil fuels at the current rate impact the environment?
3. What are the alternatives to the continued use of fossil fuel energy sources?
4. What are adverse health effects of current energy usage?
5. How would solar energy benefit society?
6. How can solar energy be used to benefit the economy?

REFERENCES

[1] N.S. Lewis, and D.G. Nocera, "Powering the planet: chemical challenges in solar energy utilization", *Proc. Natl. Acad. Sci. USA,* vol. 103, no. 43, pp. 15729-15735, 2006. [http://dx.doi.org/10.1073/pnas.0603395103] [PMID: 17043226]

[2] FACTBOX: CSP and Photovoltaic Solar Power, "Reuters", *Available: http://www.reuters.com/article/2009/08/24/us-energy-maghreb-desertec-sb-idUSTRE57N01720090824?sp=true.*Accessed: 27-Jun-2015 [Online]

[3] J. Houghton, Y. Ding, D. Griggs, M. Noguer, P. van der Linden, X. Dai, and K. Maskell, *Climate Change 2001-IPCC Third Assessment Report: The Scientific Basis.* 2001. Intergovernmental Panel on Climate Change (IPCC)

[4] "Global Environment Outlook 3", In: *Earthscan Publications Ltd for and on behalf of the United Nations Environment Programme.* 2002.

[5] *World Energy Assessment Energy and the Challenge of Sustainability*.. Bureau for Development Policy, United Nations Development Programme, 2000.

[6] R. Philips, *Energy as an Instrument for Socio-Economic Development*.. United Nations Development Programme, 1995.

[7] B. Dale, Growing Poor Slowly: Why We Must Have Renewable Energy, "OurEnergyPolicy", *Available: http://www.ourenergypolicy.org/growing-poor-slowly-why-we-must-have-renewable-energy/*.[Accessed: 27-Jun-2015] [Online]

CHAPTER 4

Solar Powered Technologies

Sarah Mousa[2,#], **Mohamed El-Morsi**[1] and **Salah El-Haggar**[1,*]

[1] *The American University in Cairo, Department of Mechanical Engineering, Cairo, Egypt*

[2] *Arab Studies and Development, 5 Fox Glove Court, Wynantskill 12198, New York, USA*

Abstract: Solar energy is a viable alternative to traditional energy sources it is healthier for both the environment and humanity, and has the potential to enhance the social and economic fabric of a community. In this chapter the basic mechanisms of several solar powered technologies are addressed. These technologies include passive building design, solar dryers, solar cookers, solar water desalination, solar water heating, solar electricity, and solar fuels.

Keywords: Biofuels, Evacuated tube, Flat plate, Humidification-dehumidification, Hydrogen, Overhangs, Passive solar building, Photosynthesis, Photovoltaic, Reverse osmosis, Solar cooker, Solar desalination, Solar dryers, Solar fuels, Solar still, Tent dryer, Thermal chimney, Thermal mass, Trombe Wall, Trough.

INTRODUCTION

This chapter provides a brief overview of selected solar powered technologies, including their mechanisms and possible applications. As a continuation for the previous chapter, this chapter provides a brief overview of selected solar powered technologies, including their mechanisms possible applications and their potential benefits. This prepares the reader and furnishes the floor for the following chapter where the implementation of these different technologies is being presented.

* **Corresponding author Salah El-Haggar:** Department of Mechanical Engineering, School of Sciences and Engineering, The American University in Cairo, AUC Avenue P.O. Box 74, New Cairo 11835, Egypt; Email: elhaggar@aucegypt.edu.

Founder of Shamsina

Salah El-Haggar, Sarah Mousa & Mohamed El-Morsi (Eds.)
All rights reserved-© 2016 Bentham Science Publishers

Chapter 5 will feature specific case studies involving the technologies discussed here.

PASSIVE SOLAR BUILDING DESIGN

The objective of passive solar building design is to allow sunshine into a building when heat and light are needed and to prevent heat from entering during warmer periods. If such designs can be integrated into the structure of a building, the need for energy sources to maintain moderate temperatures can be significantly reduced. Using passive solar design techniques make buildings more energy-efficient and allow for less reliance on traditional energy sources for lighting, heating and cooling. Passive solar is optimal when designed into a building's original structure; making alterations to a structure can be much more difficult [1].

There are a number of features which can be integrated into a building design to make it more energy efficient. The first requisite for a passive solar building is proper orientation. The longest side of the building should face south so that maximum surface area is exposed to the sun, thus allowing for the greatest amount of light to enter during the day. Second, windows and skylights as well as outdoor trees should be strategically placed. The appropriate placement of trees can result in significant reductions in heating or cooling costs. With the appropriate placing of skylights and windows, a good solar building should not require active lights before sunset. Furthermore, window sizes and glazing (double glazing, for example, helps retain heat) should be taken into consideration. The third consideration should be shading and overhangs to reduce solar gain during summer months but permit it during winter months. Fourth, thermal mass should be used in construction to allow for the maximum storage of heat during winter months. Building materials such as concrete, tile, bricks and stones have a lot of thermal mass and if incorporated into a structure can reduce the need for heating during winter [2].

To minimize the need for cooling, for hot regions like the Middle East and the Arab world, high albedo building surfaces can be used to reduce solar gains, buildings can be oriented to catch the prevailing wind direction for wind-driven natural ventilation. Also, urban designs that minimizes the heat island effect can be adopted. This can be achieved through considerable shading *via* overhangs, awnings and vegetation, and anchoring buildings thermally to cooler ground using deep foundations and tile/stone flooring. Additionally, ground source heat exchange can be used to provide additional low energy cost cooling

The solar path and its seasonal variation, as discussed in Chapter 1, must be taken into consideration when determining window orientations in a solar building. Fig. (**4.1**) features a passive solar building. As shown in the figure, the orientation of windows and panels accounts for the seasonal variation in the suns zenith so that sunlight can enter during the winter, but can be shielded during the summer [3].

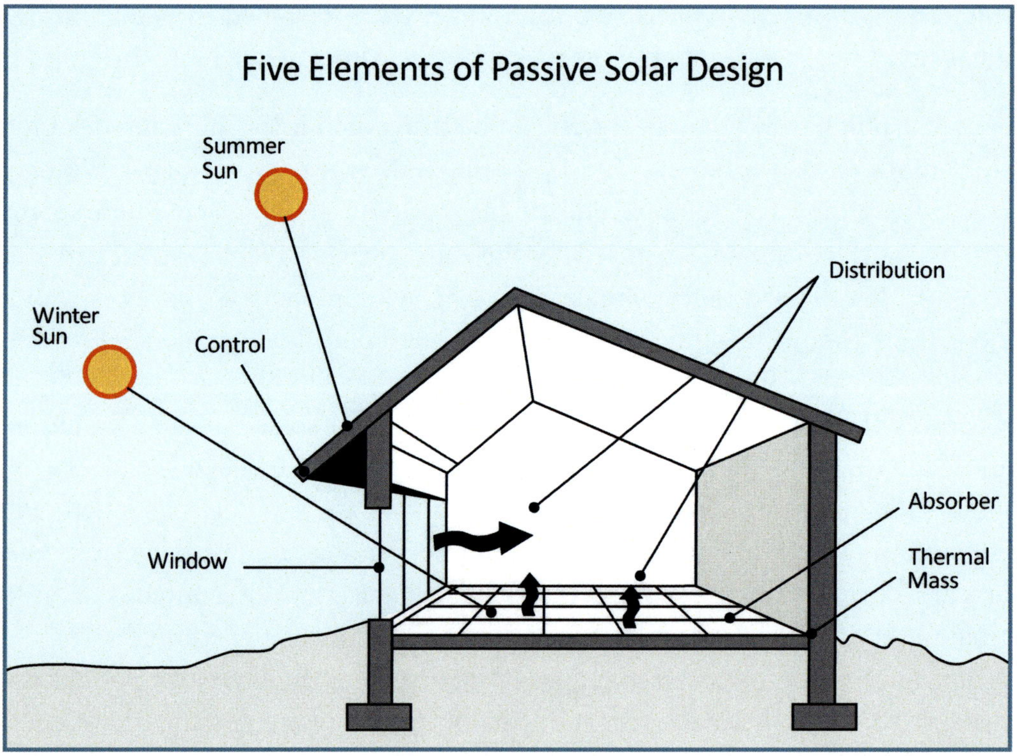

Fig. (4.1). Five elements of passive solar design [3].

A passive solar building can also include a ventilation mechanism referred to as a solar or thermal chimney. In its simplest form, it is a chimney that is painted black to attract solar energy. During the day, the hot air fills the chimney to create a suction effect (convection) that ventilates and cools the air below. The chimney must be located on the side of the house facing the sun. In addition to a solar chimney, ducts can also be located at the bottom of a building to allow cool air in during summer months. Fig. (**4.2**) shows the chimney, as well as ducts to allow cool air to enter the building.

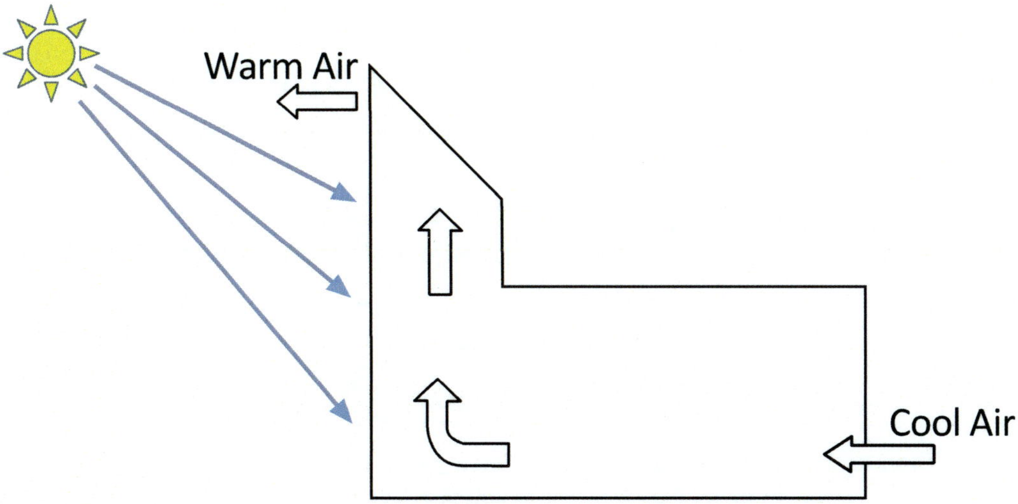

Fig. (4.2). Solar chimney and duct.

An additional ventilation method which relies upon similar principles is referred to as the Trombe Wall. This wall can be used both for heating and cooling purposes depending on the season or time of day. The wall consists of two layers: a wall of thermal mass at the interior of the building, and an outer glass wall that is separated by air space. The inner wall features vents at the top and bottom, and is painted black on the side that faces the glass to attract solar energy. The sun heats the air in between the wall and glass layer, thus creating a convection effect that draws cool air out of the interior of the building and allows warm air to enter [2]. Fig. (**4.3**) illustrates a Trombe wall.

The designs discussed are basic mechanisms of solar passive design. The specific designs selected for a building largely depend on the needs of a particular geographical location. For example, an efficient building in a location that is particularly cold will be very different from one in a hot climate. Furthermore, materials available and historical influences can also have an impact on the specific form that a passive solar building in a particular location will take. Chapter 5 will further elaborate on this notion with the case studies it will present from the different parts of the world.

SOLAR DRYERS

Solar dryers are primarily useful for the drying of agricultural products, ranging from rice to fruits to timber. Agricultural products often need to be dried either for preservation for later use, as in the case with several foods and especially fruits, or as an intrinsic part of the manufacturing process such as for wood or tobacco. The main aim of the drying process is to lower the amount of moisture in a product to a targeted value. For thousands of years, people have used the sun (which causes water content to evaporate) and wind (which blows evaporated moisture from the surface of the product) in open air to dry agricultural products.

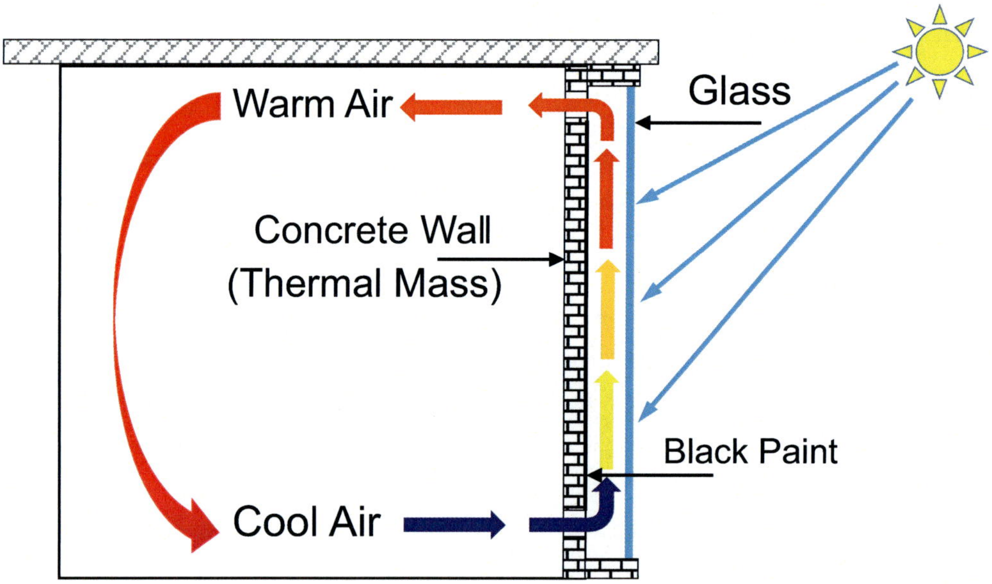

Fig. (4.3). Trombe wall.

In many developing countries, methods of open-air drying still persist. In developed countries, this methodology has largely been replaced by mechanized dryers, which use boilers for air heating and fans to move the hot air, along with moisture released from the product, through an enclosed space. Mechanized drying is quicker than open-air drying and often produces an enhanced product due its more precise temperature control. There are drawbacks however, as the equipment needed is both expensive and consumes considerable quantities of energy to operate. Solar dryers serve as a viable alternative to both of the two more prominent methods of drying, open air and mechanized drying. Solar dryers are more effective than open-air drying and less costly than mechanized dryers on several levels.

Traditional open-air drying is often accomplished by spreading a crop on the ground, perhaps on an open sheet, and exposing it to the drying effects of the sun and wind. The sun's heat evaporates the moisture in the product, while wind works to remove the evaporated moisture. Open-air drying remains the most widely recognized technique for drying in tropical developing countries. It remains attractive because of its low costs. In Egypt, as in many Arab countries, rice harvest season in villages is coupled with the sight of rice arranged on large sheets on the ground. While open-air drying does achieve the main goal, it has several drawbacks, including risk of insect infestation, and contamination with dirt. Additionally, crops can be inadequately dried due to a lack of control over temperatures or weather.

The main goal of a solar drying system is to concentrate heat on a product in order to transfer the humidity from the center of a product to its surface and thus from the surface to the encompassing air. There are two major types of solar drying systems (*i*) passive solar energy dryers (natural convection dryers) and (*ii*) active solar energy dryers (hybrid systems). Active dryers, or hybrid dryers, involve the combined use of solar and other energy sources.

Fig. (**4.4**) features a common design for a passive solar dryer. It is designed particularly for rice, but its basic principles hold for other products and solar dryer designs. In the design above, cool air is convected naturally in the upward direction by buoyancy. As it passes through the collector, the air is heated by the sun, as the surface of the tunnel is transparent and the bottom layer is dark in order to increase the absorbed energy from the sun. The air is then slightly cooled as it removes the humidity from the rice, located in the box chamber of the dryer. The chamber where the rice is located is covered by a clear material, such as glass or plastic. Thus, the incoming warm air and the sun heat the rice. The air naturally flows up the chimney by convection and is released since the cooler outside air is denser than the inside warm air in the dryer. This forces air circulation throughout the system. Circulation increases as the height of the area above the product increases. It should be noted that many products, such as rice or beans, are damaged by excessive temperatures and it may be necessary to consider temperature control depending on the local climate. The same basic principles discussed are displayed in Fig. (**4.5**) in an alternative solar dryer design, where drying of a various materials takes place simultaneously.

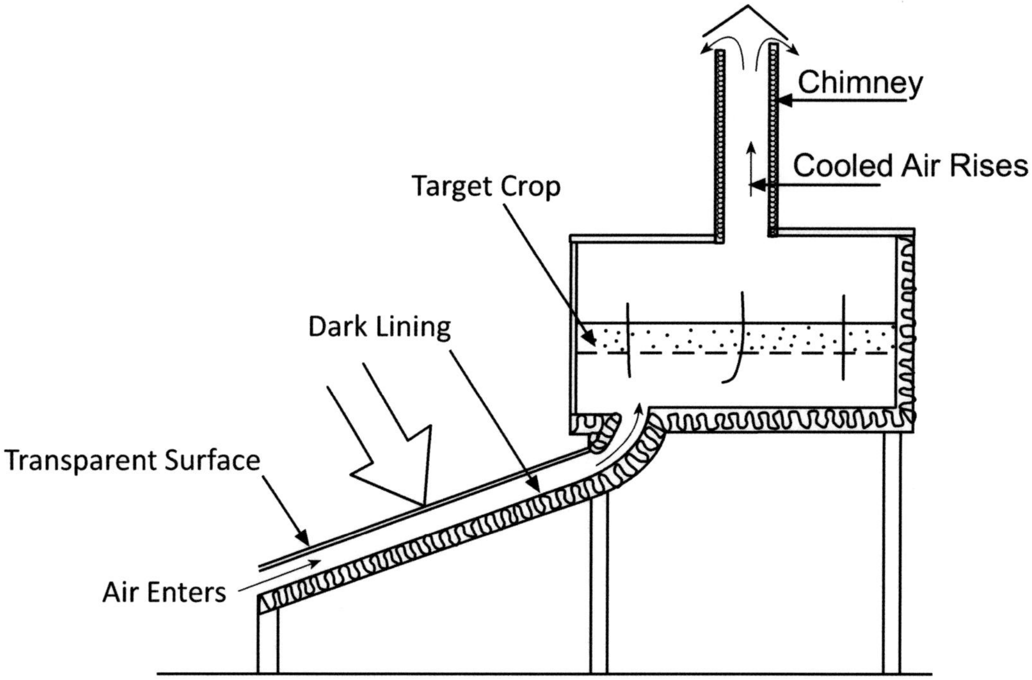

Fig. (4.4). Passive distributed (indirect) type solar dryer [4].

Other types of solar dryers include the forced convection dryer and the tent dryer. In a forced convection dryer, airflow is created using a fan. This reduces the drying time by three fold and decreasing the required area for the collector by half. This drying form, however, is considered a hybrid system since it does not solely rely on solar energy. Depending on the type of energy used to power the fan, this dryer form could have some environmental costs.

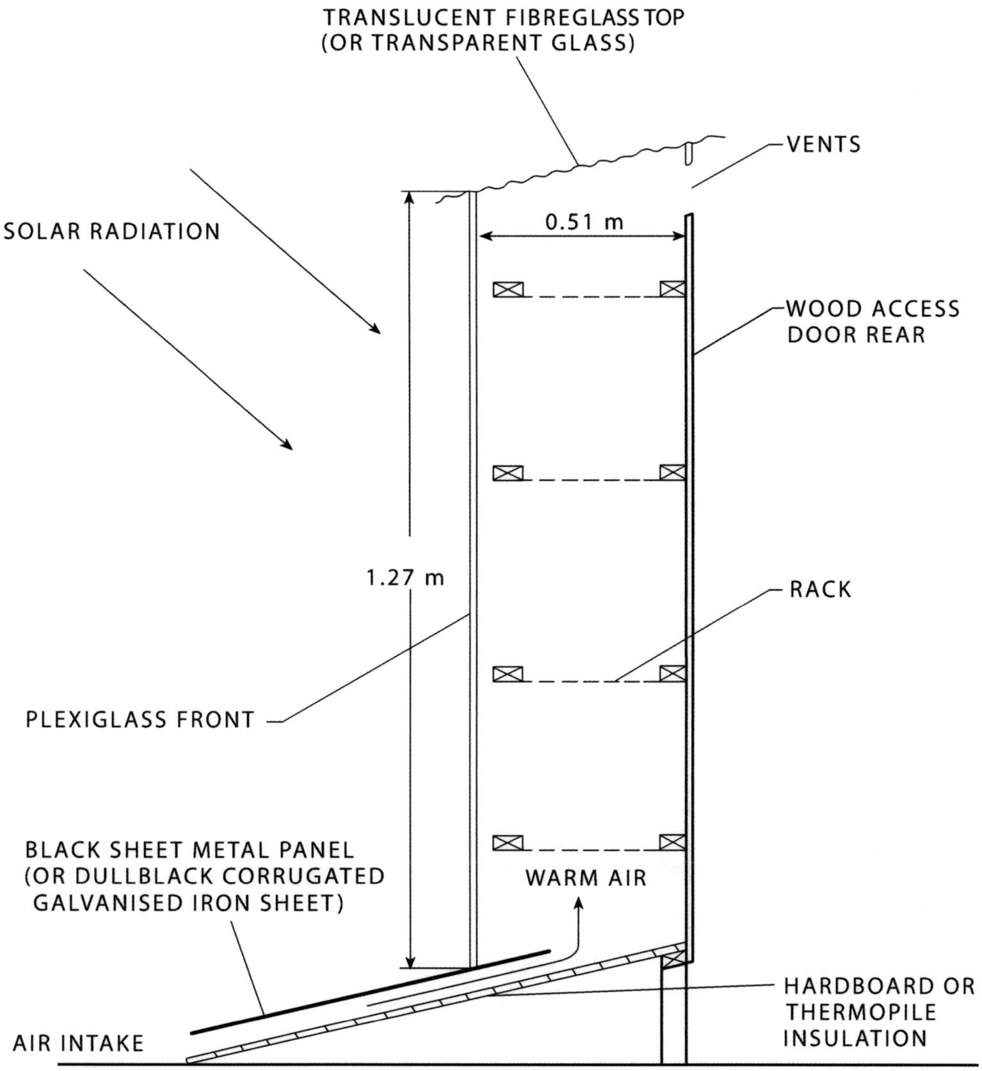

Fig. (4.5). Multi-stacked mixed-mode natural-circulation solar-energy dryer [4].

A tent dryer, the simplest form of the solar dryer, costs less to construct than previously discussed types of dryers and reduces much of the risks for insect or dirt contamination. The drying time, however, is nearly the same as open-air drying. Tent dryers consist of a large mat upon which the food is placed and a tent covered by transparent material. At the bottom of the tent, there are openings to allow air to enter. At the top of the tent, there should also be an opening to allow air to exit, as shown in Fig. (**4.6**).

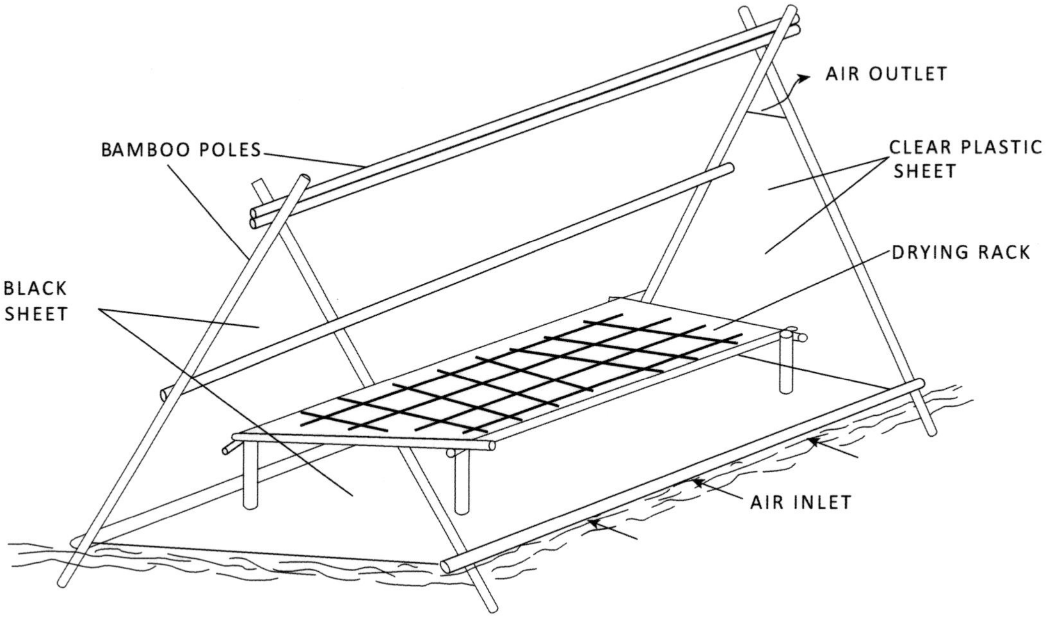

Fig. (4.6). Natural-circulation polythene-tent dryer [4].

An important advantage of solar drying as opposed to open-air drying is the protection of the agricultural product from the weather, insects, animals and dust. Faster drying also decreases the possibility of mold growth. Higher drying temperatures associated with solar drying systems allow for more complete drying, and this may allow for extended storage potential.

One of the biggest drawbacks associated with passive solar powered dryers is a lack of control over the temperature. Temperatures reached in a solar dryer can be inconsistent throughout the day and differ greatly from one season to another. Many crops are sensitive to the temperature and require a particular amount of heat to dry. One possible solution to this problem is a hybrid solar dryer, which relies on a combination of solar power and another energy source to provide a consistent temperature throughout the day. Such a system consumes less energy than a purely mechanical dryer would while maintaining consistent temperatures [5].

In many developing countries, where open-air drying is common, solar dryers are necessary not only to speed up the drying process, but also to help ensure that crops are not contaminated by insects or dirt. At minimum, tent dryers can be used in place of open-air drying for crops to create a more sanitary process. Tent dryers are easier to construct than other types of solar dryers and should be feasible for even the most underdeveloped communities. More efficient types of dryers such as those featured in Figs. (**4.4** and **4.5**) may be more difficult to build but cost less than mechanized dryers and should also be explored as an alternative to current methods of open-air drying where feasible.

SOLAR COOKERS

The fundamental purpose of a solar cooker is to heat food using energy from the sun. There are numerous types of solar cookers, with varying costs, quality of materials and efficiency [6]. Solar cookers are especially important in remote or off-grid areas where energy is unavailable. As discussed in Chapter 2, a large portion of the world's population does not have regular access to electricity and uses wood as a source of fuel, resulting in adverse effects to the health of both the environment and people [7].

An efficient solar cooker maximizes heat gain and minimizes heat loss to allow for the maximum absorption and retention of heat. There are several factors which must be accounted for when constructing a solar cooker. First, the surface layer must be able to absorb and retain the maximum amount of heat. Second, the materials of the cooker itself must also share these qualities. Additionally, durability must be taken into account. A solar cooker should consist primarily of an enclosed space covered by a clear material, such as glass or plastic, which allows for sunlight to enter. Sunlight light that passes, to the enclosed space, through the cover is absorbed and reflected by materials within the space. Dark shades absorb the sunlight to a higher degree, and should be elected for both the interior of the cooker and any pots or materials used in the cooking process [8].

The glass or plastic covering of the cooker should face the sun as directly as possible in order to maximize heat gain. At noon, east/west dimension of the box should be longer to optimize exposure to solar energy for the greatest amount of time. The glass itself may also be tilted to more directly face the sun. In Fig. (**4.7**), the glass in box 1 and box 2 are identical, yet, more sun radiates through the glass in box 2 because of its optimal orientation. (Box 2, however, also has more interior surface area through which heat can be lost) [8].

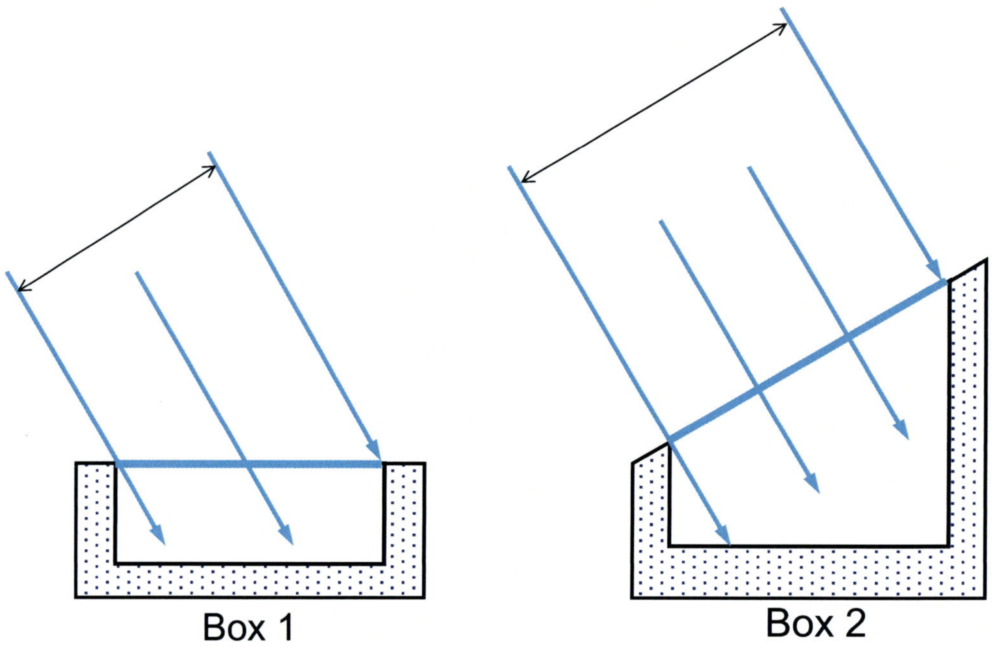

Fig. (4.7). Different glass orientations in a solar box cooker [8].

Another important factor to consider in the construction of a solar cooker is the type of material used for the surface. Double-glazing is considered optimal for solar cookers due to its higher retention of heat. One drawback is that, depending on the transmittance of cover material, heat gain may be reduced. Single or multiple reflectors can be added to a solar cooker for additional heat gain. Reflectors reflects extra sunlight through the cover into the solar box, resulting in higher cooking temperatures. The orientation of the reflectors is important, and it should be noted that multiple reflectors run the risk of blocking sunlight. Fig. (**4.8**) shows possible reflector orientations [8].

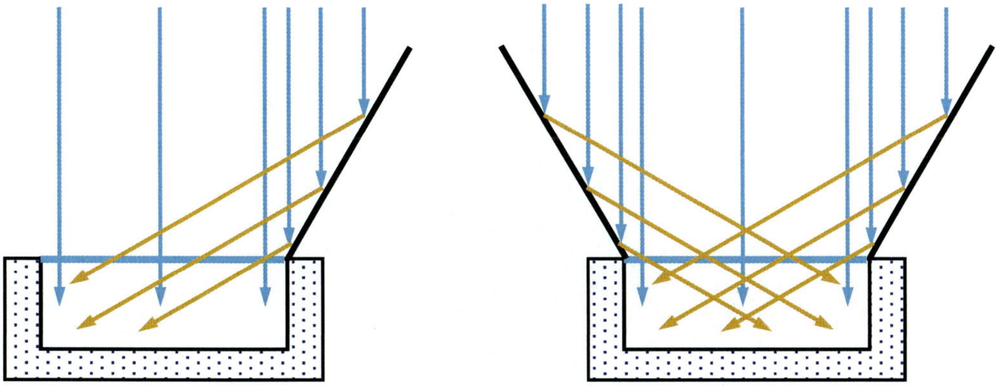

Fig. (4.8). Possible reflectors orientations for solar box cookers [8].

Thermal energy (heat) always travels from hotter to cooler surfaces. Heat within a solar box cooker can by conduction, radiation, and convection. Conduction is when heat travels through materials, as illustrated in Fig. (**4.9**), where the heat from a pan passes through its handle. Similarly, the thermal energy inside a box is lost when it passes through the box construction materials such as tin foil, glass, cardboard, inside air and insulation, to the outside air [8].

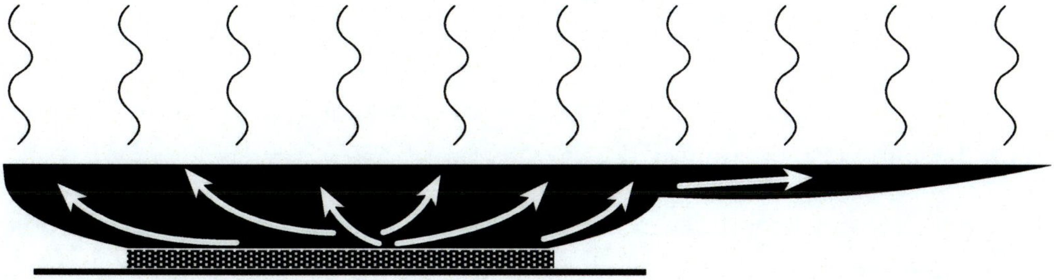

Fig. (4.9). Heat conducted through the pan to handle [8].

There are ways to minimize conduction in a solar box cooker. One way involves the metal absorber plate at the bottom of a cooker. This plate conducts heat from the pot to the materials which it touches at the bottom of the cooker. Small insulating spacers are inserted between the absorber plate and the bottom of the box to minimize conduction losses, as shown in Fig. (**4.10**). This reduces the surface area of materials which the absorber plate touches [8].

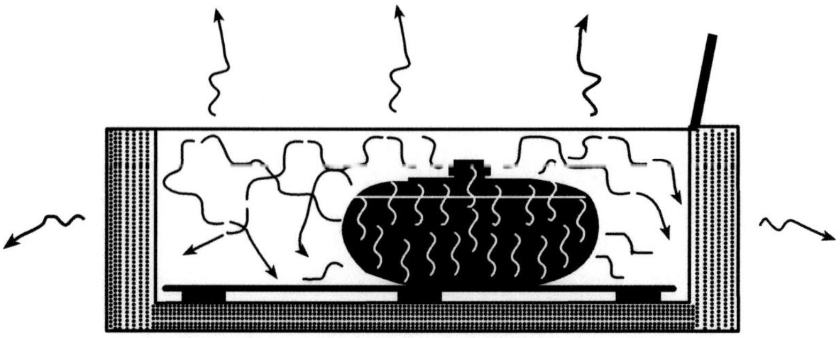

Fig. (4.10). Minimizing conduction by using small insulating spacers [8].

The second method for heat loss, radiation, occurs when materials that are warm or hot radiate heat to their surroundings through air or vacuum. The foil inside a solar box reflects back most of the energy radiated from the hot pots and the bottom tray. Since glass and plastics are transparent to short-wave radiation, but opaque to long-wave radiation, most of the energy radiated from the interior of the solar cooker is trapped inside by the cover. However, plastics are not as efficient in trapping radiant energy as glass.

The third method for heat loss, convection, occurs when air molecules travel in and out of a solar cooker box through cracks. Warm air from the interior of the solar box can escape through the cracks around the top cover or construction imperfections. Also, the cooler outside air leaks from the outside to the interior of the box through these openings [8].

As the density and specific heat of the cooker materials increase, the cooker becomes capable of holding more heat, as shown in Fig. (**4.11**). There is, however, a drawback to using denser materials; the interior of a box including heavy materials such as rocks, bricks, heavy pans, or heavy foods will take longer to heat because of this additional heat storage capacity. The incoming energy is stored as heat in these heavy materials, slowing down the heating of the air in the box. These dense materials, with the increased heat which they store, radiate that heat within the box, keeping it warm for a longer period of time [8].

Structural materials are important to consider for the durability and transportability of a solar cooker—these materials give the cooker its shape and form and determine how long it will survive. Wood, metal, bricks, glass, bamboo, cement, stone, plastic, cardboard, paper mache, clay and much more are potential structural materials. Most materials which perform well structurally are too dense to perform well as insulators [9]. To reach high enough temperatures for cooking, all walls and bottom, should be well insulators. Aluminum foil, feathers, rice hulls, wool, straw, or newspaper are examples for good insulation materials that can be used in a solar box cooker, except for the cover that should be made from a transparent material [9].

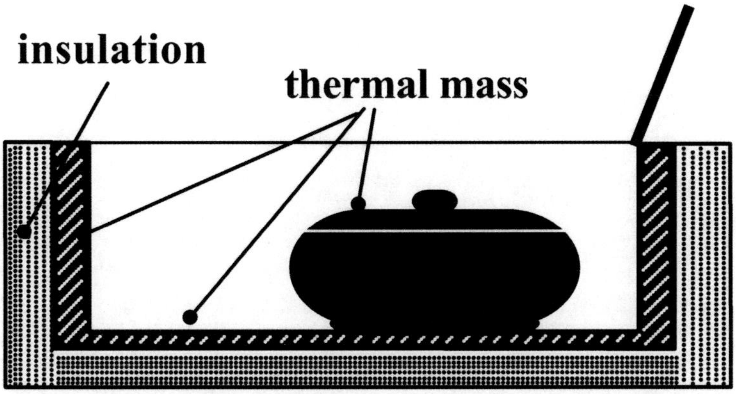

Fig. (4.11). Thermal mass in solar box cookers [8].

In the solar box, when food containing water is heated, the created vapor pressure drives the moisture from the inside to the outside of the box. If there is no barrier for moisture, it can directly escape through gaps and cracks in the box or it can be forced into the walls and the bottom of the box. Water vapor could be trapped if the box is designed with both high quality seals and moisture barriers, inside the cooking chamber. While designing the majority of solar box cookers, it is necessary that the inner-most surface of the cooker is a good vapor barrier. By slowing the water vapor movement into both the bottom and the walls of the cooker, this vapor barrier will prevent water damage to structural materials and insulation [8].

The size of a solar box cooker should take into consideration the quantity of food usually cooked. Second, when considering size the utility of a box should also be considered: if a cooker will be used in different areas it should be easily transportable. And third, a box design must be spacious enough for commonly used cookware. Box cookers, parabolic and panel cookers are the most common three solar cookers designs. There are many variations to these three types of solar cookers, but they represent the basic structures [6]. The three main types of cookers are illustrated below in Figs. (**4.12** to **4.14**). The parabolic cooker, in Fig. (**4.13**), is more efficient than the box cooker, as the curved surface allows for greater concentration of energy. A panel cooker combines design concepts from both the box and parabolic cooker. It is more effective than the box cooker, but more easily transportable and less costly than the parabolic cooker [6]. Selection of a specific cooker will depend on the specific needs and preferences of the user regarding price, efficiency and transportability.

Fig. (4.12). Solar box cooker [8].

SOLAR WATER DESALINATION

Limited fresh water sources makes desalination necessary for ensuring a sustainable water supply for human consumption and irrigation. The need for fresh water is especially critical in more arid parts of the world, where desalination processes have been used increasingly in recent years. The process of desalination is not a modern one, but modern technology has made it more efficient. Nonetheless, desalination is expensive and requires great quantities of energy. The energy costs required for desalination are especially significant for water-short countries, many of which cannot afford the required fossil fuel resources [10].

Fig. (4.13). Parabolic cooker [8].

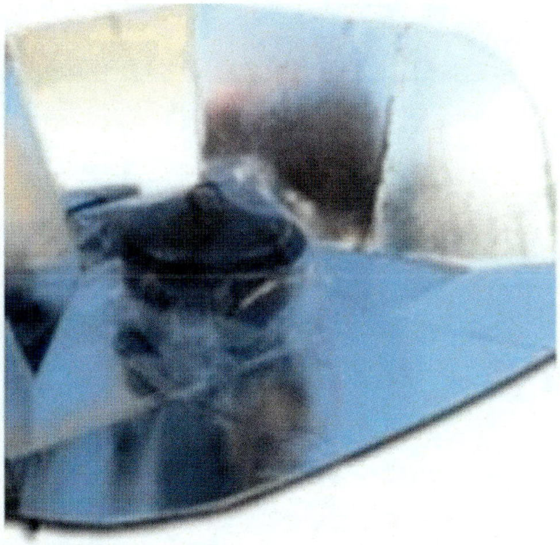

Fig. (4.14). Panel cooker [8].

There are two basic methods for water desalination: membrane and thermal desalination. Several technologies powered either by traditional energy sources or by solar energy sources are derived from these two basic methods. Be electing to use solar energy for desalination, a large part of the operational costs involved in the process can be eliminated. Many arid areas which are characterized by water shortages, are also wealthy in solar energy sources, thus providing great potential for solving the dilemma [10].

One of the two basic processes, membrane desalination, for desalinating water containing salts or minerals involves reverse osmosis—a pressure-driven system that pumps salt water through a semi-permeable membrane. Reverse osmosis is illustrated in Fig. (**4.15**), where salt water is forced through a semi-permeable membrane that prevents salts or minerals from crossing the barrier [11].

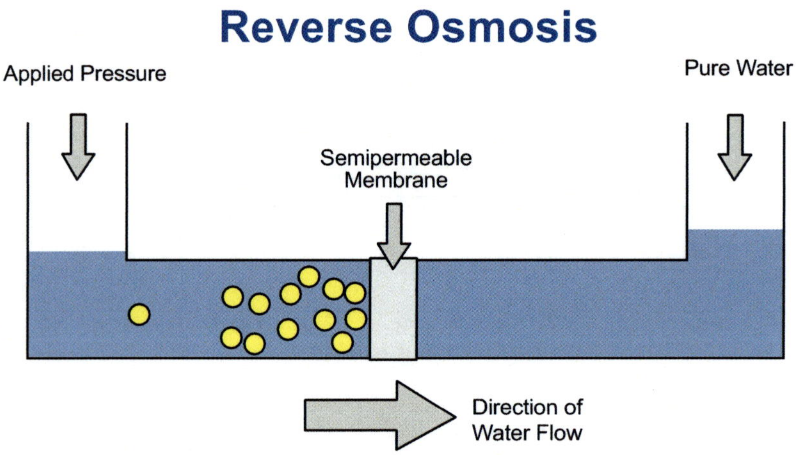

Fig. (4.15). Reverse osmosis [11].

In most currently used systems, fossil fuel energy sources are relied upon to adhere pressure forcing salt water across a membrane. In recent years, systems which rely upon solar energy for this purpose have been developed. In Fig. (**4.16**), photovoltaic panels power the pump which drives salt water through a membrane and into a storage tank for fresh water. The system is portable and has been suggested for use in emergency situations when traditional energy sources are not easily accessible [12].

The second type of desalination system is the thermal desalination. In its basic form, the system relies on thermal energy to evaporate salt water, separating it from salts and minerals, and then condensing the evaporated fresh water. Solar energy can be used to evaporate salt water, which then condenses in a separate, cooler chamber. This system is relatively inexpensive and is useful for small-scale desalination. It occurs in nature when sea water is evaporated by solar energy, condenses, and falls as rain in the form of fresh water [10]. Fig. (**4.17**) illustrates a solar humidification-dehumidification still in its simplest form; it is referred to as a single-effect solar still [10]. Solar energy is concentrated on the salt water, which evaporates, condenses as freshwater at the top of the still, and falls into a water collector.

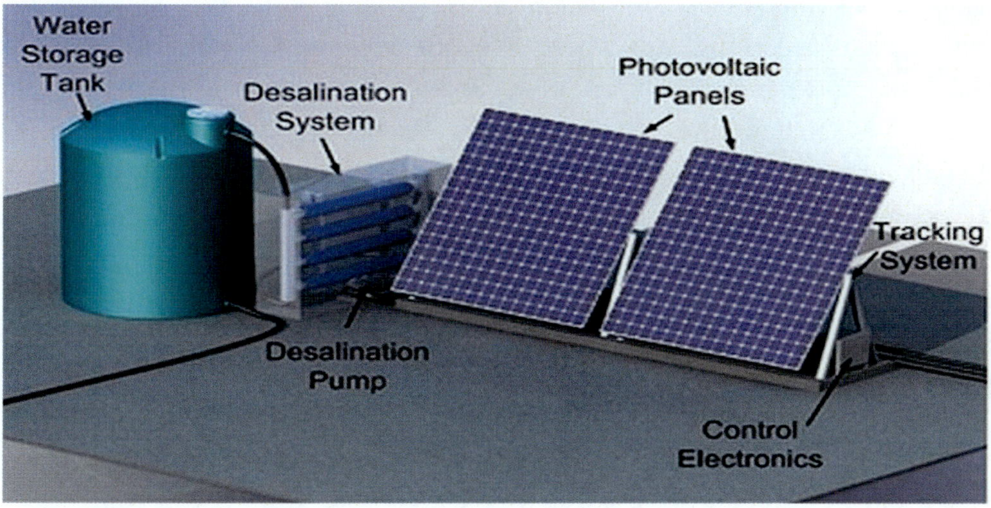

Fig. (4.16). Solar driven reverse osmosis desalination system [12].

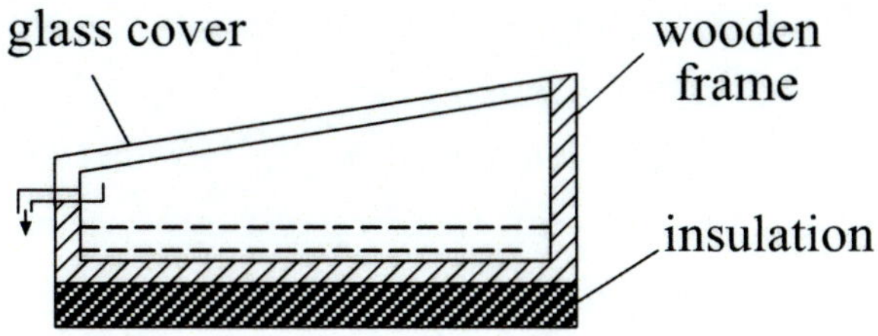

Fig. (4.17). Solar Humidification-Dehumidification Still [13].

To heat the water, there are several solar technologies that can be utilized such as the flat-plate collector, the evacuated tube collector, parabolic trough. Flat-plate collector and evacuated tube collector are similar to the technologies that apply solar energy to heat domestic water, to be discussed in the next section. The flat plate collector involves a set of metal pipes, painted in a dark color, attached to an insulated box. Salt water flows through the system, and is evaporated and rises into a collector tank where it condenses as fresh water. The evacuated tube collector functions on the same principle but is more efficient than the flat plate collector. The single difference between the two systems is that the evacuated tube collector relies on a more sophisticated piping design.

Fig. (**4.18**) illustrates a thermal desalination system using evacuated tube heat pipes. The heat pipes heat and evaporate salt water, and the evaporated water travels through a cooling coil where it condenses and is stored in the "water deposit" as shown below in Fig. (**4.18**). Fig. (**4.19**), shows the structure and principle of evacuated tube heat pipe solar water heater. Inside of the outer glass tube is a heat pipe, which is a sealed hollow copper tube that contains a selected liquid, which boils under low pressure and low temperature. The liquid boils and travels up the tube as a vapor, transferring the heat to a header located inside of the heating barrel. The heat is absorbed by the header, and the liquid condenses and returns to the base of the heat pipe, and the process continues [14].

An additional desalination system is the salinity-gradient solar pond. This pond contains salt water, which naturally forms a vertical salinity gradient. The heavier salts and minerals remain at the bottom of the pond, while fresh water can be found at the top layers of the pond. The density of the salts increases with the depth. A solar pond has three main layers: a bottom layer with a high concentration of salt, an intermediate layer which serves as a gradient and a top layer with low concentrations of salts. As solar energy is focused on the pond, the bottom layer absorbs this energy. Freshwater from the bottom layer will rise to the top, but the salts at the bottom layer do not rise due to their higher density. The solar pond, in addition to desalination, thus serves a second purpose due to its storage of energy. A natural example of a solar pond can be found in Egypt's Sinai Peninsula [10].

There are several other solar desalination systems, as well as variations of the basic systems described above. The systems discussed, however, illustrate the main concepts that operate solar desalination systems. While desalination systems powered by solar energy are currently scarce, there are great prospects for the systems described above. Solar desalination systems cost less to run than those powered by fossil fuels and are a viable alternative to economically and environmentally costly methods [16].

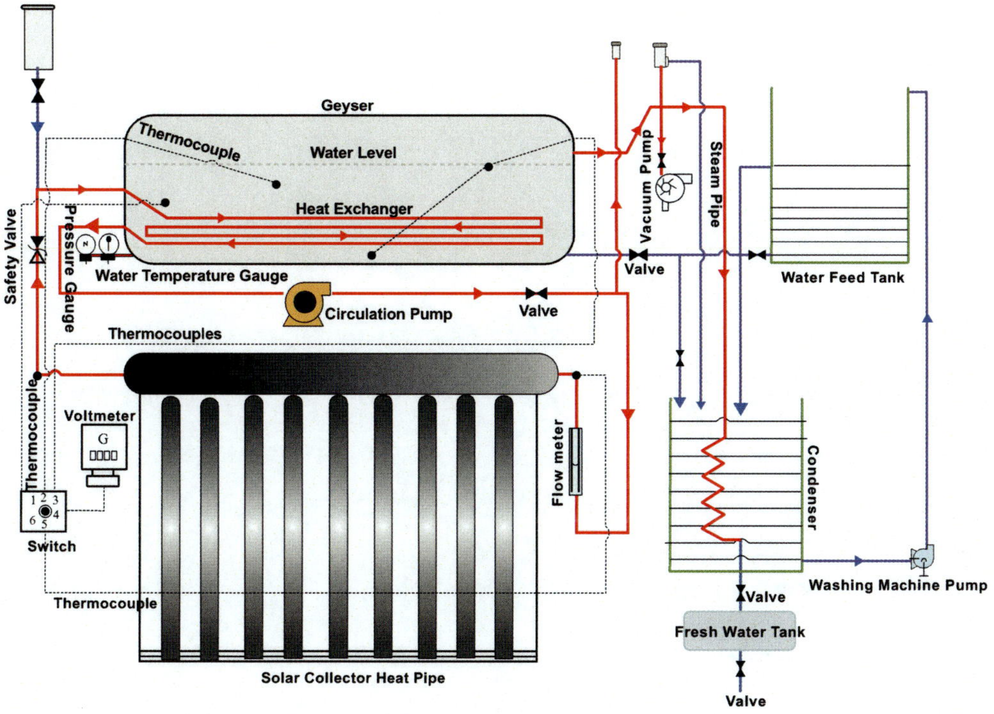

Fig. (4.18). Evacuated tube water desalination system [15].

SOLAR POWERED WATER HEATERS

Hot water is a basic humanitarian necessity and is used residentially for a variety of purposes, including cooking, cleaning, bathing, and heating. By using solar energy to heat water, this basic necessity can be provided without the environmentally damaging effects of fossil fuel powered heaters. Solar powered water heaters are of particular importance to communities without access to modern energy sources [17].

Water heaters are currently most commonly powered *via* electricity, gas or oil.

The use of these power resources, as discussed in Chapter 2, is associated with adverse effects on the environment, and can be costly for consumers. In some poor or rural areas where such power resources are unavailable, water is heated by burning rubbish or other materials. According to a report released by the United Nations Development Program (UNDP), the installation of solar powered water heaters can significantly reduce greenhouse gas emissions [18].

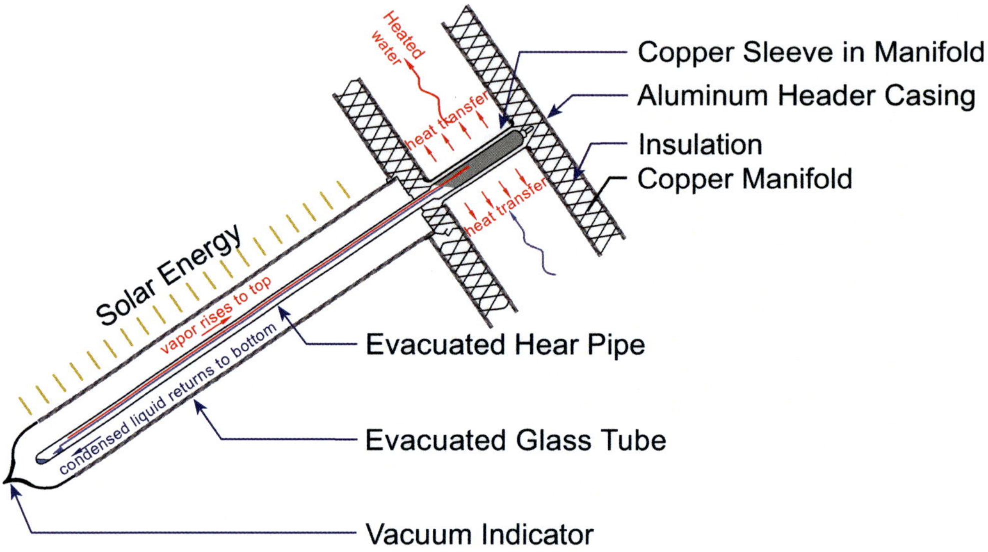

Fig. (4.19). Structure of evacuated tube heat pipe solar water heater [14].

The concept of a solar-powered water heater is not contemporary; early models date back to at least the late 19th century. There are two main types of solar-powered water heating systems: flat panel solar water heaters and evacuated tube solar water heaters. Flat panels or flat plate solar collectors are currently the most common type used for solar water heating in many countries. For each of these two types of water heaters, there are also active and passive heaters. An active water heater requires a pump that forces water through the system, while a passive system works *via* natural convection and is not reliant on other energy sources.

There are numerous designs for flat panel solar-powered water heaters. A common design primarily consists of a durable, weatherproof box, lined with a dark absorber plate into which copper tubes are welded, and topped with a transparent (usually glass) cover. Fig. (**4.20**) features the main components of a flat plate solar water heater.

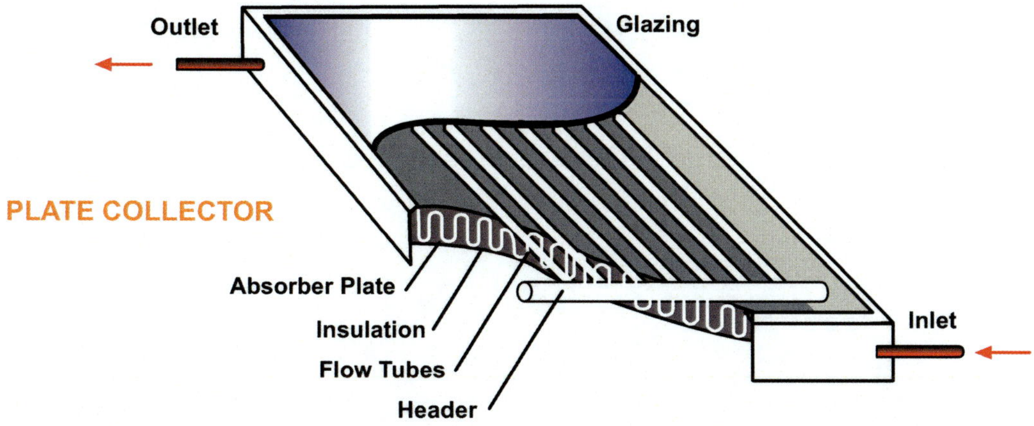

Fig. (4.20). Main components of a flat plate solar water heater [19].

Flat panels are associated with several shortcomings, however, which have led to the development of an improved type of solar water heater which features evacuated tubes. Drawbacks of the flat panel system include heat loss due to the air gap between the copper tubes and the glass cover. Flat plate systems are also relatively difficult to install and maintain—any breakage in a part of the system usually necessitates replacement of the system in its entirety. Flat panels also directly face the sun only at midday, thereby decreasing their efficiency throughout later parts of the day.

The second main type of solar powered water heater features evacuated heat tubes. Evacuated heat tubes used for solar powered water heaters are identical to those described in the previous section on water desalination. The tubes are constructed of strengthened (borosilicate) glass. The outer layer of the tube is a double layer. The outermost layer is completely transparent to allow for the entry of sunlight. The inner part of this outer layer can be either transparent, as shown in Fig. (**4.21a**), or selectively coated to allow for energy absorption without reflection, as shown in Fig. (**4.21b**). These glass layers are fused at a high temperature; air is pumped out of the empty space which remains between the layers to decrease heat loss.

There are several features of the evacuated tube system which make it superior to the flat plate system. As mentioned previously, the lack of air space significantly decreases heat loss and provides insulation. The use of a liquid which boils at low temperatures also hastens the heating process. Evacuated tubes are also relatively strong. Any damage to a tube will not affect the rest of the system, and tubes can be individually replaced. Furthermore, because evacuated tubes are round, they passively track the sun throughout the day, providing heat for a longer time. Fig. (**4.22**) compares the pattern by which flat panel and evacuated tube systems track the sun throughout a day.

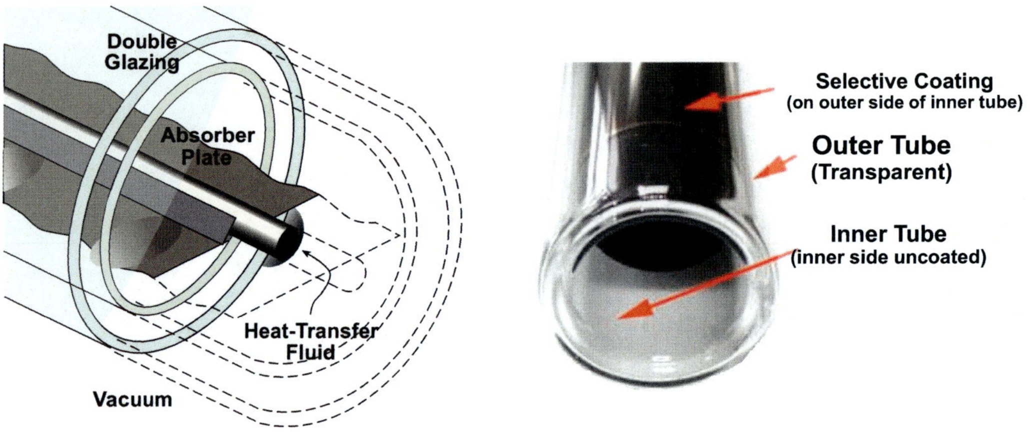

(a) [20] (b) [21]

Fig. (4.21). Construction of evacuated tube.

The Incidence Angle Modifier (IAM) is the variance in the solar heat absorption of a solar collector as the angle of the sun changes throughout the day. When the system is perpendicular to the sun, the maximum IAM value reached is equal to one. In flat-plate collectors, this value of one is achieved at midday when the sun is directly above the system, while the value is lower throughout other parts of the day. In contrast, evacuated solar tubes reach values exceeding one during the morning and afternoon periods, when the solar path is shorter than at other times of day, as the cylindrical design allows for enhanced performance. Fig. (**4.23**) shows IAM values for a flat panel and for an evacuated tube system designed by

Solar Panel Plus (SPP).

While the evacuated heat tube solar powered water heaters have clear advantages over the flat panel system, flat panels remain more prominent. Reasons for this may be associated with the lower cost and simpler design of flat panels. While efficiency is important, both of these factors make flat panels more practical in many cases.

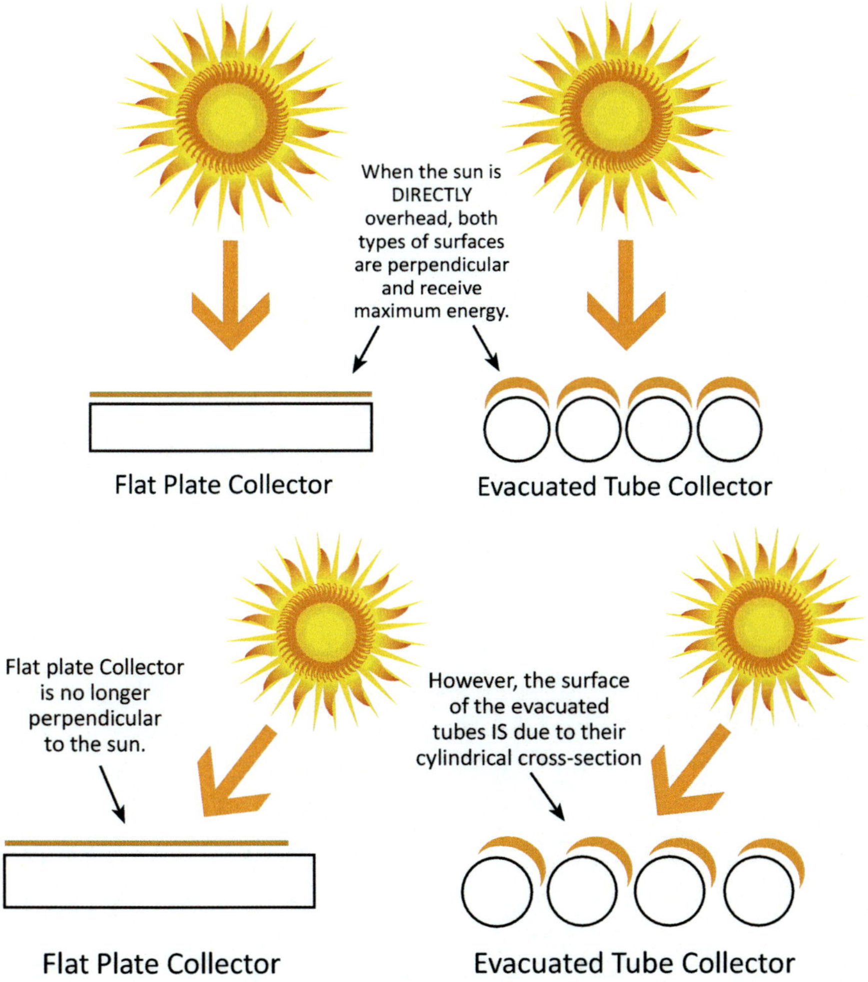

Fig. (4.22). Passive tracking in flat panels and evacuated tubes [22].

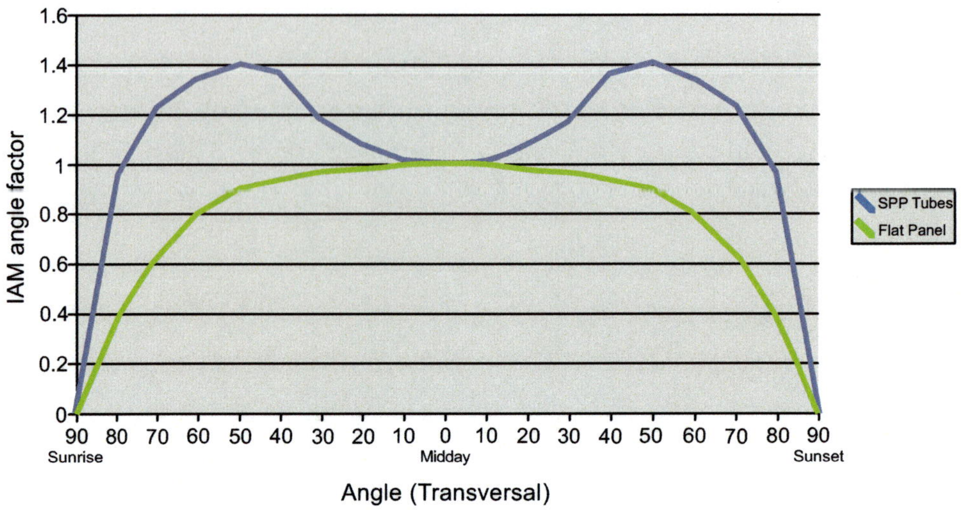

Fig. (4.23). Solar collector IAM curves [23].

SOLAR ELECTRICITY

The technologies discussed in previous sections of this chapter have all relied upon garnering solar energy and converting it into useable heat. This section of the chapter will focus on technology used to convert solar energy into electricity. There are two processes through which sunlight can be converted into electricity: the photovoltaic (PV) solar and thermal solar processes.

Electricity can be produced from solar energy *via* the PV effect, which relies on a series of PV cells which comprise a solar panel. The PV effect is created by combining two layers of semi-conductive materials, usually silicon. The wafers (thin slices) of silicon are cut into segments, and are doped (treated with a dopant) to create an electron imbalance. One of these layers has an excess number of electrons and the other a depleted number of electrons. The negatively charged layer is referred to as the N-Type. The second layer, which has a shortage in electrons, is referred to as P-Type. A thin layer is placed between the two layers so that electricity does not flow between them without the presence of sunlight.

When sunlight shines on the layers, photons from the light are absorbed by the material and excite the electrons causing them to jump from one layer to another.

This process generates an electric current, as shown in Fig. (**4.24**). The wafers are then aligned to make a solar cell and conductive metal strips are used to frame the cells and carry the generated electric current. The system is topped with an anti-reflective layer and a glass layer to allow for the maximum absorption of sunlight [24].

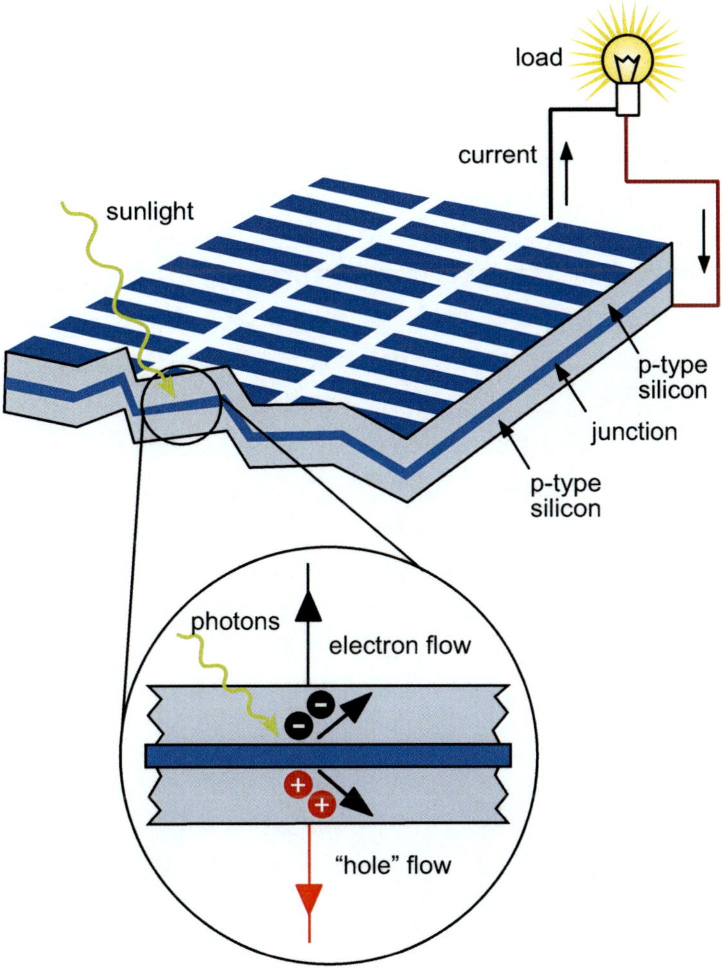

Fig. (4.24). Photovoltaic solar panel [25].

Solar electricity can often be more costly if other sources of energy are available, but it is ideal for remote or off-grid locations. Solar electricity can also be important in areas which experience power cuts or shortages, and in cases of emergency. Solar electricity can be useful for generating electricity for lights or smaller appliances, yet the quantity of solar cells needed to generate electricity for more electrically demanding processes is not practical, although further research and development in the field is expected make the process more practical in the future.

While the installation of PV solar panels can be costly, once installed it is a virtually free source of energy that does not cause damage to the environment. Maintenance required for the system once installed is relatively low. The manufacturing process does leave a carbon footprint, partly due to chemicals required for the doping process, but overtime this carbon footprint has been significantly reduced.

A second way to produce solar electricity is through a thermal process. Through this process, the energy of the sun is used to heat a liquid and generate steam. Curved mirrors or lenses (a solar thermal trough) are used to concentrate sunlight, heating liquid that creates steam which turns a turbine. This parabolic trough consists of curved lenses/mirrors which focus sunlight on a glass tube containing a liquid that runs the entire length of the trough, as shown in Fig. (**4.25**). Some are mobile, designed with a tracking system that follows the sun's movement, tilting the trough east and west so that it faces the sun more directly.

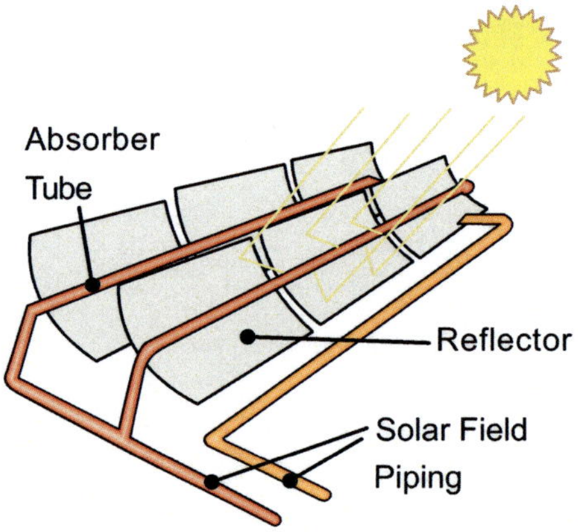

Fig. (4.25). Thermal solar electricity trough [26].

The glass tube containing fluid (receiver) can also be enclosed in a glass vacuum chamber, reducing heat loss. The easily heated fluid (such as synthetic oil or molten salt) is transported to a heat engine where it is transferred into electricity.

A heat engine, shown in Fig. (**4.26**) works by taking advantage of the temperature difference between a hot fluid or gas and a cold sink. The difference in temperature turns the turbine, indirectly creating electricity [27].

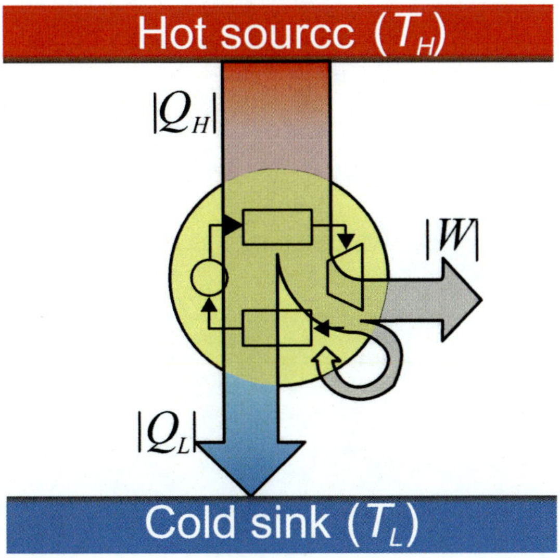

Fig. (4.26). Photosynthesis [29].

SOLAR FUELS

The motivation for the production of solar fuels on large scale in the past few years has been a field of interest for the global research community [29]. Energy from the sun can be stored in chemical bonds of materials. The stored energy can be released whenever needed [30]. These materials carriers chemical energy, also called solar fuels. These fuels can be easily transported and/or stored from one place to the other. Solar fuels can either be (*i*) hydrogen that can be used in fuel cells for electricity production, or (*ii*) carbon-based fuels that can used in fuel cells to produce electricity, where a synthesis gas from carbon monoxide and hydrogen is used, or (*iii*) hydrocarbon fuels "biofuels" that can be burned to release thermal energy that can be converted to electricity.

Hydrogen is produced *via* solar electrochemical path. Electricity from photovoltaics or solar thermal systems is used in an electrolytic process to produce hydrogen and oxygen from water, as shown in Fig. (**4.27**) [29].

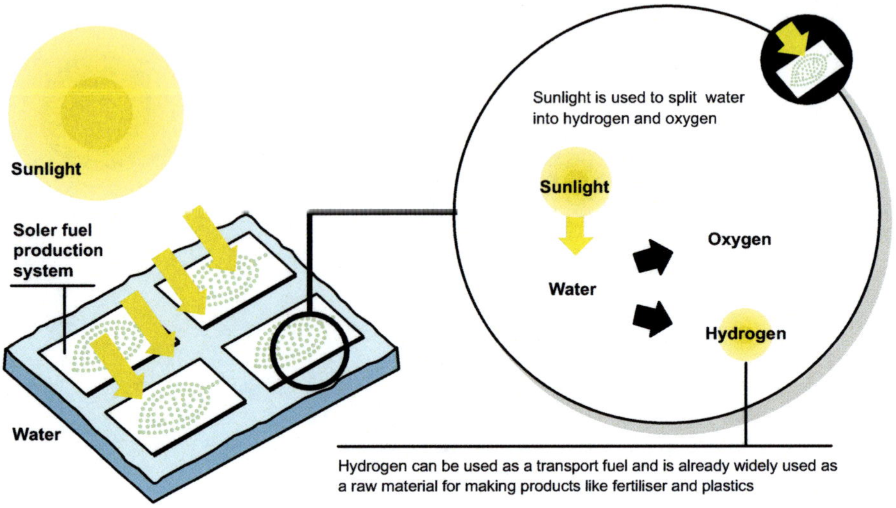

Fig. (4.27). Producing hydrogen by splitting water using sunlight [29].

Carbon-based fuels are produced *via* solar thermochemical path. Reactants are heated to a very high temperature, using solar energy, to produce carbon monoxide and hydrogen for fuel production [29].

Biofuels are produced *via* solar photochemical natural path. In nature energy from the sun is used in the conversion of both water and carbon dioxide, by photosynthesis, to oxygen and carbohydrates, as shown in Fig. (**4.28**). The conversion efficiency of solar energy to chemical energy, by natural photosynthesis, is less than 1%. However, algae has a conversion efficiency of 5-10%, cyanobacteria is also more efficient. Conversely, artificial photosynthesis can has a maximum of 40% conversion efficiency [30].

Since the beginning of the 20th century, intense focus has been placed on cultivating crops to produce first generation biofuels, but with time, newer problems have emerged. First generation biofuels rely primarily on food crops such as corn, wheat, palm oil and the like. This caused a never-ending debate on the topic of "food *versus* biofuel" land use, and the increase in food prices due to diverting the use of many of the grains to the biofuel production instead of the human food chain. This paved the way for second and third generation biofuels to emerge.

The third generation biofuel consists mainly of using algae to generate biofuel. Growing algae can generate at least 8 times more gallons per acre per year and 40 times more gallons/acre/year compared to palm oil and rapeseed respectively, two of the highest oil content crops cultivated for first generation biofuel production as shown in Table **4.1**.

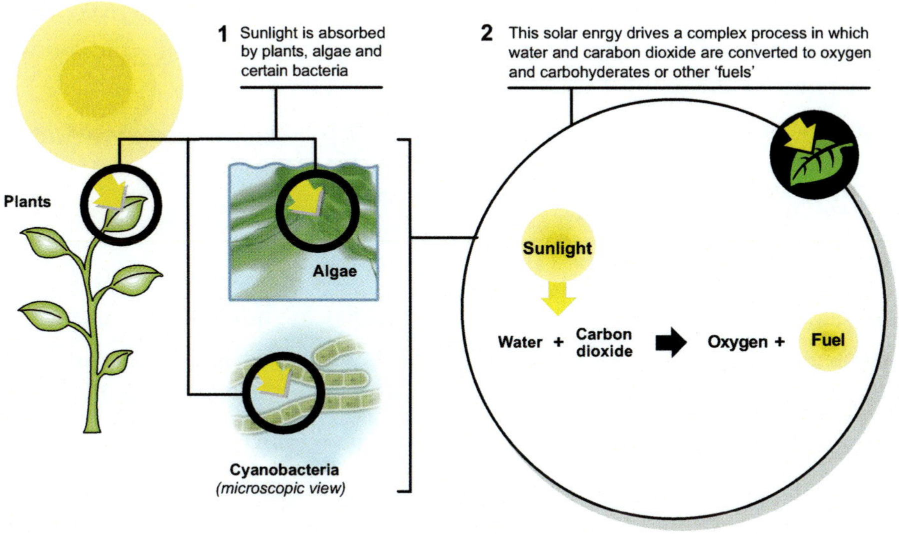

Fig. (4.28). Photosynthesis [29].

Table 4.1. Photosynthesis [31].

Crop	Oil yield (L/ha)	Land area needed (Mha)
Corn	172	1540
Soybean	446	594
Canola	1190	223
Jetropha	1892	140
Coconut	2689	99
Oil Palm	5950	45
Algae (30% oil)	58700	4.5
Algae (70% oil)	136900	2.0

Algal biofuel is considered an innovative source for renewable/bioenergy resource. The oil in microalgae can be converted into jet fuel or diesel fuel.

Microalgae are highly productive, but cannot be used on agricultural land or products, and are carbon-neutral. Algae are aquatic organisms that have similar characteristic to plants, but have no roots. Algae can grow in salt water, fresh water, even wastewater, and in a variety of different temperatures. Algae can literally grow in any water body such as ponds, lakes, aquarium tanks, *etc.*

In order to live, these plant-resembling organisms rely on photosynthesis, in which light, CO_2, water and nutrients (primarily nitrogen and phosphorus) are used to synthesize algae food (in form of glucose) and release oxygen as a waste product. Algae then use glucose to produce fats (lipids and oil), starch and protein.

DISCUSSION

This chapter has addressed basic mechanisms of several solar powered technologies. Most of the technologies discussed, in their simplest form, can be easily constructed from accessible and inexpensive materials. When selecting and designing a specific technology for construction, it is important to note the basic design features discussed in this chapter and their purposes. While it is possible to alter these designs, this should be done in the framework of maintaining or enhancing the purpose of maximizing energy absorption and retention. Chapter 5 will present case studies of each of these technologies; while the details of the technologies may differ from the ones presented here, all maintain the core design features which contribute to the appropriate harnessing of solar energy.

QUESTIONS

1. What are some design features of a passive solar building?
2. What are the benefits of a solar dryer? In what communities would solar driers be ideal?
3. What communities would benefit most from solar cookers? How might the design of a solar cooker vary depending on the types of families or communities which use it?
4. Why is solar water desalination necessary? Why do Arab Mediterranean countries have particular potential in the usage of the solar energy for water desalination?
5. In the context of solar water heaters, why are evacuated tubes more effective

than flat panel collectors?

6. What are the advantages of solar electricity as opposed to traditional electricity sources?

REFERENCES

[1] Windows, Doors, & Skylights, "US Department of Energy", *Available: http://energy.gov/ public-services/homes/windows-doors-skylights.*Accessed: 24-Jun-2015 [Online]

[2] Passive Solar Home Design, "US Department of Energy", *Available: http://energy.gov/ energysaver/articles/passive-solar-home-design.*Accessed: 24-Jun-2015 [Online]

[3] "Guide to Passive Solar Home Design, "US Department of Energy", *Available: http://energy.gov/ sites/prod/files/guide_to_passive_solar_home_design.pdf.*Accessed: 24-Jun-2015 [Online]

[4] O. Ekechukwu, and B. Norton, "Review of Solar-Energy Drying Systems II: An Overview of Solar Drying Technology", *Energy Convers. Manage.,* vol. 40, no. 6, pp. 615-655, 1999. [http://dx.doi.org/10.1016/S0196-8904(98)00093-4]

[5] N.M. Khattab, "Optimization of Hybrid Solar Dryer", *Energy Sources,* vol. 18, no. 7, pp. 781-790, 1996. [http://dx.doi.org/10.1080/00908319608908810]

[6] Solar Cooker Designs, "Solar Cookers International Network (SCInet)", *Available: http://solarcooking.wikia.com/wiki/Category:Solar_cooker_designs.*Accessed: 25-Jun-2015 [Online]

[7] Solar Cookers International, Annual Report Highlights, "Solar Cookers International (SCI)", *Available: http://www.solarcookers.org/report/summary/.*Accessed: 25-Jun-2015 [Online]

[8] M. Aalfs, Principles of Solar Box Cooker Design, "Solar Cookers International Network (SCInet)", *Available: http://solarcooking.wikia.com/wiki/Principles_of_Solar_Box_Cooker_Design.*Accessed: 14-Jun-2015 [Online]

[9] Equipment Needed and the Rationale, *Available: https://whatssizzling.wordpress.com/2008/03/30/ equipment-needed-and-the-rationale/.*Accessed: 25-Jun-2015 [Online]

[10] H.M. Qiblawey, and F. Banat, "Solar Thermal Desalination Technologies", *Desalination,* vol. 220, no. 1–3, pp. 633-644, 2008. [http://dx.doi.org/10.1016/j.desal.2007.01.059]

[11] Reverse Osmosis, Hydrocomponents & Technologies Inc., "Hydrocomponents & Technologies Inc", *Available: http://www.hcti.com/sm/aboutro/aboutro.html.*Accessed: 25-Jun-2015 [Online]

[12] "MIT Unveils Portable Solar-Powered Water Desalination System, *Available: http://inhabitat.com/ mit-unveils-compact-solar-powered-water-desalination-system/.*Accessed: 25-Jun-2015 [Online]

[13] G. Xiao, X. Wang, M. Ni, F. Wang, W. Zhu, Z. Luo, and K. Cen, "A review on solar stills for brine desalination", *Appl. Energy,* vol. 103, pp. 642-652, 2013. [http://dx.doi.org/10.1016/j.apenergy.2012.10.029]

[14] How to Work Solar Water Heater - Heat Pipe, *Available: http://www.prosunindia.com/ whatisheatpipe.html.*Accessed: 25-Jun-2015 [Online]

[15] A. Alwaer, and J. Gryzagoridis, "Water Desalination by Evacuated Tube Heat Pipe Solar Collector", *2014 International Conference on the Eleventh industrial and Commercial Use of Energy,* 2014pp. 1-5 [http://dx.doi.org/10.1109/ICUE.2014.6904207]

[16] L. García-Rodríguez, "Seawater Desalination Driven by Renewable Energies: A Review", *Desalination,* vol. 143, no. 2, pp. 103-113, 2002. [http://dx.doi.org/10.1016/S0011-9164(02)00232-1]

[17] The History of Solar Power, "US Department of Energy", *Available: http://www1.eere.energy.gov/solar/pdfs/solar_timeline.pdf.*Accessed: 25-Jun-2015 [Online]

[18] E. Ebrahimian, *"Community Action to Address Climate Change: Case Studies Linking Sustainable Energy Use with Improved Livelihoods," Global Environment Facility (GEF).*. Small Grants Programme, United Nations Development Programme, 2003.

[19] Solar Thermal, "New Mexico Solar and Wind", *Available: http://www.newmexicosolar-andwind.com/Solar Thermal.html.*Accessed: 25-Jun-2015 [Online]

[20] Evacuated-Tube Solar Thermal Collectors, Home Power Magazine, *Available: http://www.homepower.com/articles/solar-water-heating/equipment-products/flat-plate-evacuated-tube-solar-thermal-collectors.*Accessed: 25-Jun-2015 [Online]

[21] Sydney Type Tube with Inner Glass Coating, "Designing Buildings Wiki", *Available: http://www.designingbuildings.co.uk/w/images/d/dc/Sydneytube.jpg.*Accessed: 26-Jun-2015 [Online]

[22] Passive Tracking, Sahab Ji Solar, *Available: http://www.sahabjisolar.com/solar-water-heaters.html.*Accessed: 26-Jun-2015 [Online]

[23] Incident Angle Independent Thermal Collectors, Solar Panels Plus, *Available: http://www.solarpanelsplus.com/solar-tracking/.*Accessed: 26-Jun-2015 [Online]

[24] M. Boxwell, *Solar Electricity Handbook: A Simple, Practical Guide to Solar Energy : how to Design and Install Photovoltaic Solar Electric Systems.* 2012.

[25] How does a Solar PV Cell works?, "Rural Integrated Development Service-Nepal (RIDS-Nepal)", *Available: http://www.rids-nepal.org/index.php/Solar_Photo_Voltaic.html.*Accessed: 26-Jun-2015 [Online]

[26] Parabolic Trough Solar Field Technology, "US National Renewable Energy Laboratory (NREL)", *Available: http://www.nrel.gov/csp/troughnet/solar_field.html.*Accessed: 26-Jun-2015 [Online]

[27] *T. B. Johansson, H. Kelly, A. Reddy, and Robert Williams, Renewable Energy: Sources for Fuels and Electricity.*. 2nd ed Island Press, 1993.

[28] Heat Engine, "wikimedia", *Available: https://upload.wikimedia.org/wikipedia/en/a/a2/Heat_engine.png.*Accessed: 26-Jun-2015 [Online]

[29] "Solar Fuels and Artificial Photosynthesis", Royal Society of Chemistry (RSC), 2012.

[30] H. de Vriend, and R. Purchase, *Solar Fuels and Artificial Photosynthesis Science and Innovation to Change our Future Energy Options.*. BioSolar Cells, 2013.

[31] Y. Chisti, "Biodiesel from microalgae", *Biotechnol. Adv.,* vol. 25, no. 3, pp. 294-306, 2007. [http://dx.doi.org/10.1016/j.biotechadv.2007.02.001] [PMID: 17350212]

CHAPTER 5

Solar Power and Sustainable Development

Sarah Mousa[2,#], **Mohamed El-Morsi**[1] and **Salah El-Haggar**[1,*]

[1] *The American University in Cairo, Department of Mechanical Engineering, Cairo, Egypt*

[2] *Arab Studies and Development, 5 Fox Glove Court, Wynantskill 12198, New York, USA*

Abstract: The need to account for sustainable development when it comes to energy is crucial; the major energy sources in current use are unsustainable in that they harm the environment and deplete resources for current generations and especially for future ones. Significant amounts of energy can be saved if a technology used to accomplish a given task is designed efficiently. Examples include the use of (i) passive building design to meet the heating and cooling demands of buildings, and (ii) solar energy for water heating, generating electricity, water desalination and cooking. Several case studies, presented in this chapter, from different geographic locations show that several technologies can be adopted to meet a certain energy use without scarifying the environmental, social, cultural, and/or aesthetic standards.

Keywords: Arab world, BedZED, Brundtland report, Case studies, Falkenmark index, Gaza, Hybrid ventilation, Masdar, MENA, Middle East, Passive, Solar chimney, Solar dryers, Solar energy, Sustainable development, Technologies, Thermal mass, Water desalination, Water scarcity, Zero energy.

INTRODUCTION

This chapter examines ways in which the solar powered technologies, discussed in Chapter 4, can be used for sustainable development in different regions of the world with a special focus on Egypt and the Middle East. The chapter focuses on successful case studies and also reveals the potential for solar energy technologies

* **Corresponding author Salah El-Haggar:** Department of Mechanical Engineering, School of Sciences and Engineering, The American University in Cairo, AUC Avenue P.O. Box 74, New Cairo 11835, Egypt; Email: elhaggar@aucegypt.edu.

Founder of Shamsina

Salah El-Haggar, Sarah Mousa & Mohamed El-Morsi (Eds.)
All rights reserved-© 2016 Bentham Science Publishers

in the Middle East. Case studies discussed in the Middle East are applicable worldwide, but the reader should take into consideration the availability of solar energy in the Middle East, as well as the social and cultural aspects which must be taken into account for the successful implementation of solar power technologies.

The chapter individually considers how each of the technologies discussed in Chapter 4 can be locally produced and developed as a step towards building awareness and increasing local usage of renewable energy technologies as well as building self-reliance from a grass-roots level. The clean and affordable energy obtained through the use of these solar powered technologies can potentially transform livelihoods, not only among those without any source of reliable energy, but also among the entirety of a population since all can integrate at least some of these technologies into their lifestyle and will benefit from a clean environment.

SUSTAINABLE DEVELOPMENT

In its 1987 Brundtland Report, the sustainable development was defined by the UN as "*development which meets the needs of the present without compromising the ability of future generations to meet their own needs*" [1]. Over the ensuing years and decades, a more comprehensive definition of sustainable development has surfaced. In its 2005 World Summit Outcome Document, three major pillars were highlighted by the UN for sustainable development: Social development, economic development and environmental protection. Arguments subsequently arose that culture should be an integrated part in the sustainable development, leading to the recognition of a fourth pillar. Cultural diversity was deemed an integral component of societies, associated with intellectual, emotional, moral and spiritual existence, which should be protected [1].

The need to account for sustainable development when it comes to energy is crucial; the major energy sources in current use are unsustainable in that they harm the environment and deplete resources for current generations and especially for future ones. Current energy sources are finite, meaning that their current rate of use will leave future generations without the resources to meet their energy needs. Not only the primarily technical aspects of solar powered technologies had an effect on the success of the products, but also factors that have included culture, aesthetics and appropriate technology played a significant role in that success. The word "appropriate technology" is often used to refer to the technology that is tailored to achieve a certain task efficiently, at the same time meeting specific use of energy and certain cultural, environmental and/or aesthetic standards.

There are arguments that renewable energy technologies, including solar powered technologies, are not always sustainable. The construction process of solar powered technologies sometimes consumes significant amounts of energy. This is especially the case with photovoltaic panels, which are comprised of silicon cells. Some of these materials are not only expensive, but are also rare and are in potential risk of depletion [2]. While some of these arguments against the sustainability of solar powered technologies may hold some validity, it is important first to consider the long term benefits of solar powered technologies and second to note that renewable energy research is relatively nascent and that more efficient technologies are being discovered by the day.

PASSIVE BUILDING DESIGN

Passive building design is a well-known proven concept that can help in reducing the heating and cooling demands of buildings [3]. However, in metropolitan areas, noise, pollution and urban heat island effects challenges the design. Several successful case studies of passive building design have been reported from different parts of the world, examples include the science and technology museum building on the eastern edge of Hangzhou, China [4], the library and Division of Art, Design and Architecture at the Judson College in Elgin, Illinois, US, [5], the Frederick Lanchester Library at Coventry University, UK, [6], the School of Slavonic and East European Studies at University College London, UK, [7] and the Queens Building at DeMontfort University, Leicester, UK [8]. In the subsequent sections the buildings from China and the US are presented as examples for passive building designs.

The science and technology museum in China [4] is an office building with four storeys in addition to a basement that uses hybrid ventilation for space conditioning. In the natural ventilation mode, fresh air is admitted naturally through the basement into 17 vertical shafts on the outer perimeter of the building. Air then flows in the ventilation space pushing stale air to a central light well then to the designated stacks where it leaves on top of the building, as shown in Fig. (**5.1**). Concrete ceilings are left exposed providing thermal mass for night purging and to stabilize the building temperature during day time. Only when mechanical cooling is needed and/or buoyancy forces are inadequate, the hybrid ventilation is activated. In the active mode, the basement is kept airtight to avoid the loss of conditioned air to the environment. In the active summer mode, the outside fresh air is dehumidified and pre-cooled by the air handling unit, which also forces this air into every office space using control dampers. Radiant chilled ceilings supplements the cooling of air using a ground source heat pump. In the active winter mode, the incoming outside fresh air will be pre-warmed using the air handling unit. Heating of air is supplemented using warm radiant surfaces, previously used for radiant cooling in summer, using the ground source heat pump. This hybrid ventilation strategy saves about two thirds of that used by pure mechanical ventilation.

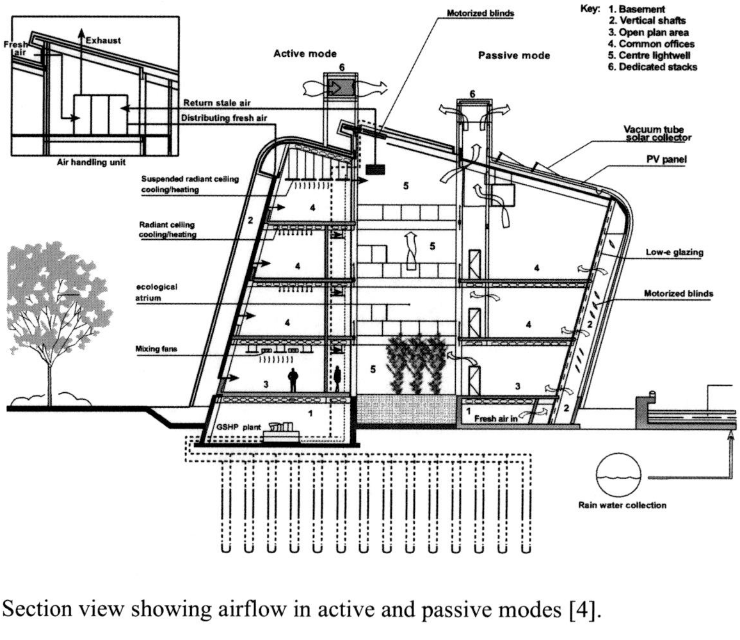

Fig. (5.1). Section view showing airflow in active and passive modes [4].

The Judson College building in the united states [5], is another example that demonstrates how buildings that combine natural ventilation with a mechanical cooling and heating system could reduce energy consumption in severe climates. The weather at this location is a severe weather that swings between -22°C and -14.8°C in January and 30°C and 35.8°C in July. The natural ventilation strategy in this building depends on delivering the fresh, from air inlet dampers at the perimeter, to the center of the library *via* a central air supply plenum. The fresh air infiltrates through the building from center-to-perimeter, where stacks around the perimeter collects the air and exhausts it to the outside, as shown in Fig. **(5.2)**.

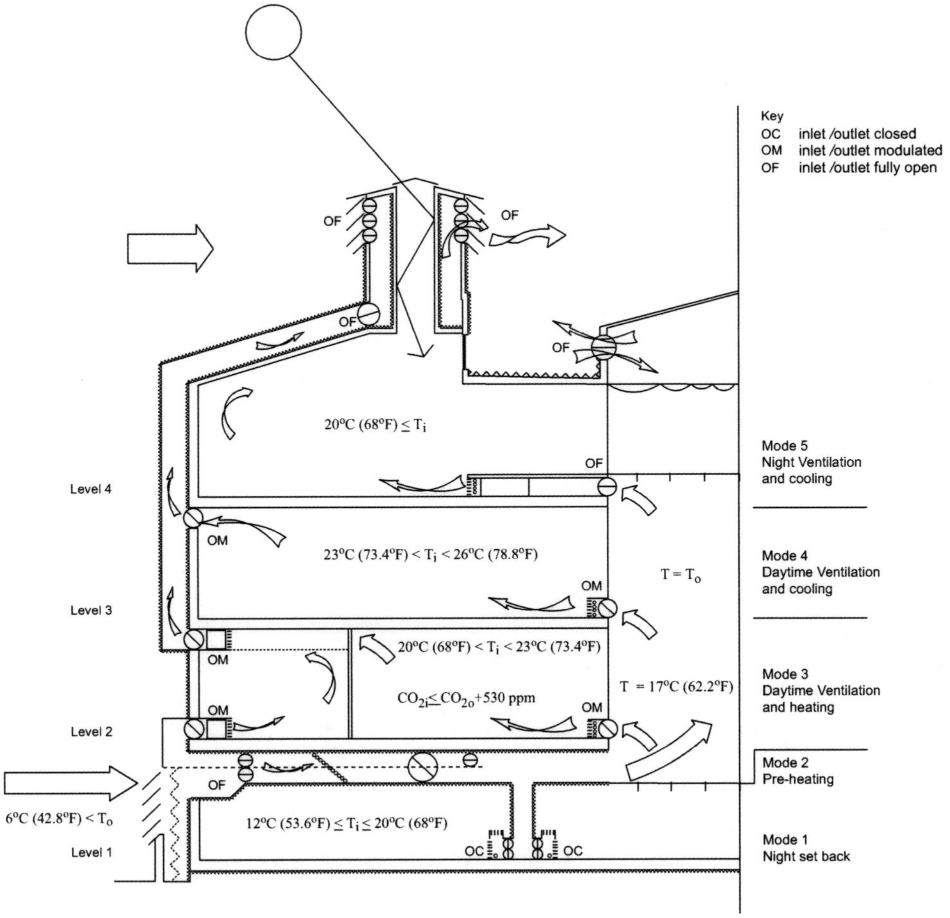

Fig. (5.2). Part section through the library during mid-season operation mode [9].

Also, concrete ceilings are left exposed providing thermal mass for night flushing and to stabilize the building temperature during day time. The air-handling unit conditions the incoming outside fresh air in both the summer and the winter. The conditioned air is then discharged into the plenum. This is done while the perimeter inlet dampers are closed, as shown in Figs. (**5.3** and **5.4**).

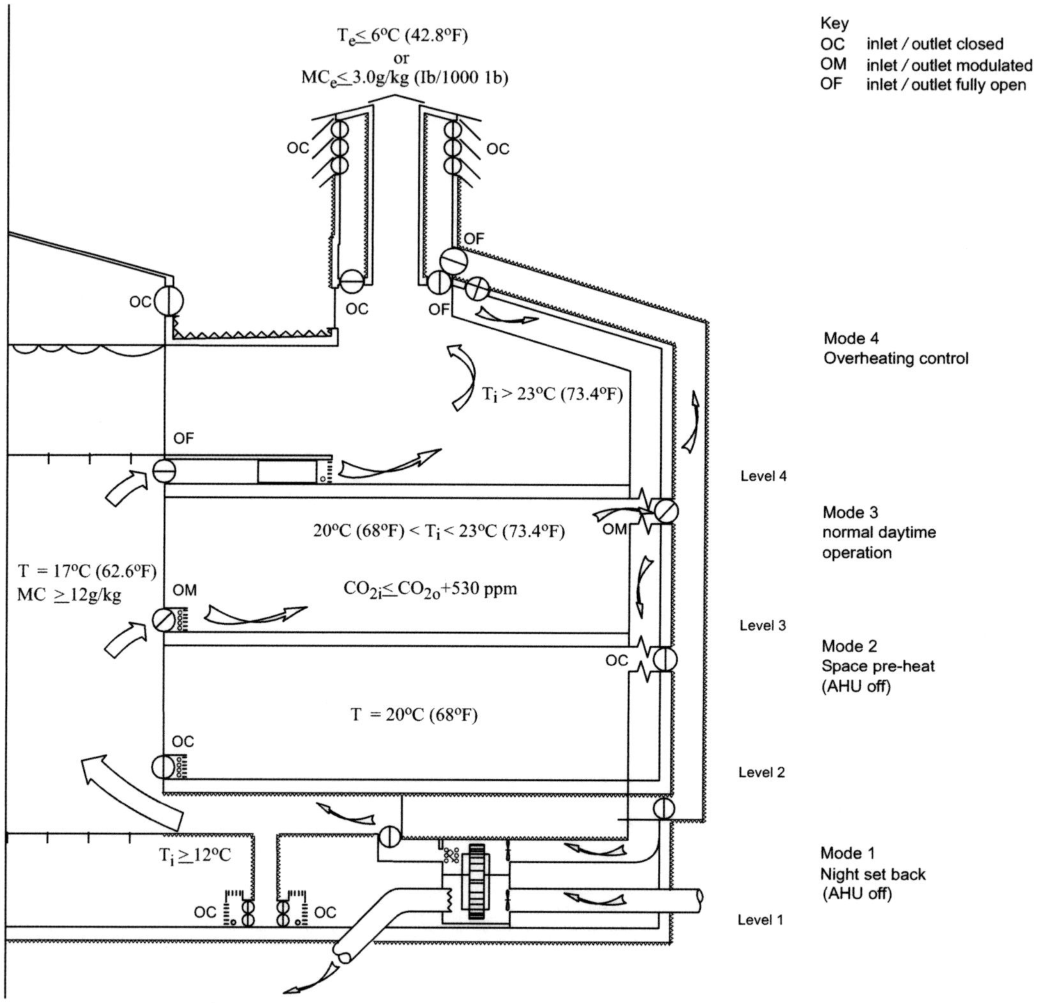

Fig. (5.3). Part section through the library during winter operation mode [9].

This hybrid ventilation strategy saves about half of that used by pure mechanical ventilation when compared against building against a standard USA building, for two cooling set points 24°C (US-24) and 26°C (US-26), as shown in Fig. (**5.5**).

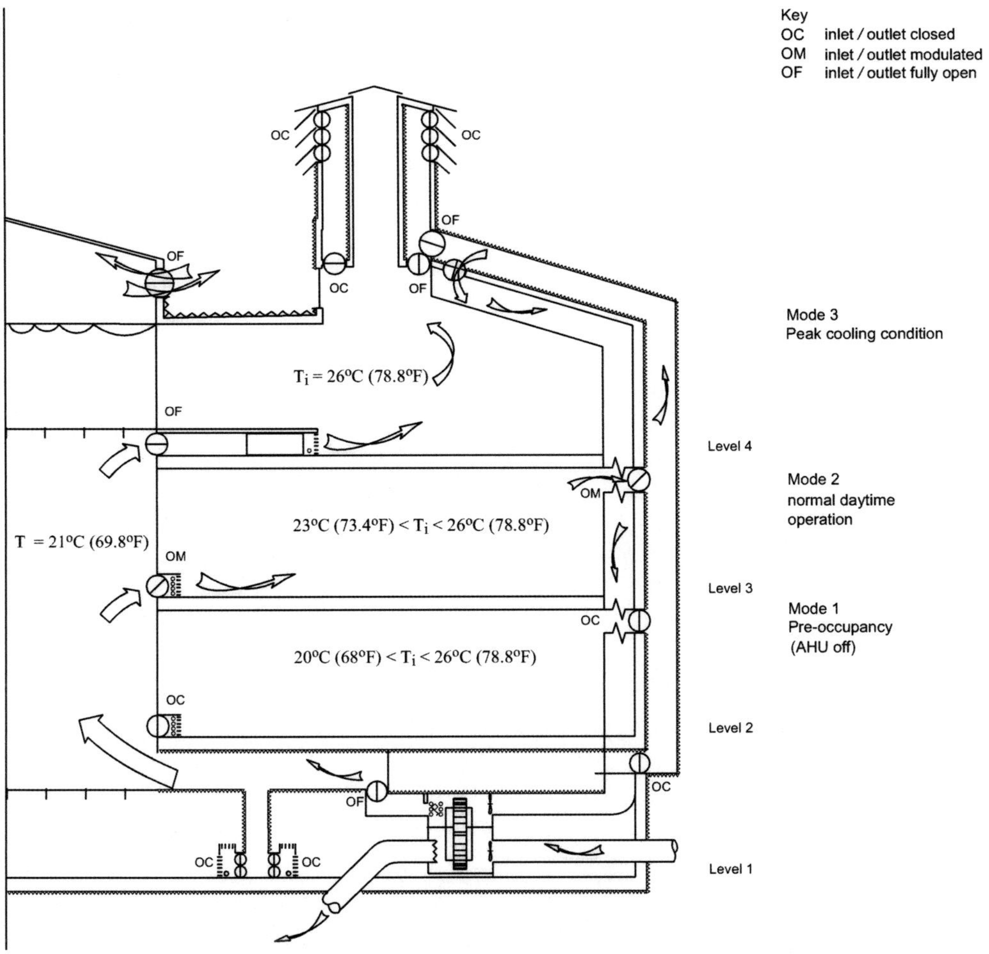

Fig. (5.4). Part section through the library during summer operation mode [9].

MVH: Mechanical ventilation and heating. MVC: Mechanical cooling.

MV: Mechanical ventilation only. PVH: Passive ventilation and heating.

PV: Passive ventilation.

When considering passive solar building design in the Arab world, it is important to take into consideration climate and cultural factors. The region's hot and arid climate, with diurnal swings (hot days and cool nights) that characterize desert climates, has provided it with a wealth of traditional architectural designs, predating modern fossil fuel-reliant technologies, which it can revive to increase energy efficiency in building designs. During the 1980s, Egyptian architect Hassan Fathy advocated aspects of traditional design with the aim of creating environmentally-responsive buildings. He has been the primary voice in this field. Fathy focused on several aspects of traditional building designs, including the courtyard and its role in controlling temperature within a building. Architects adhering to these principles have been labeled as revivalists.

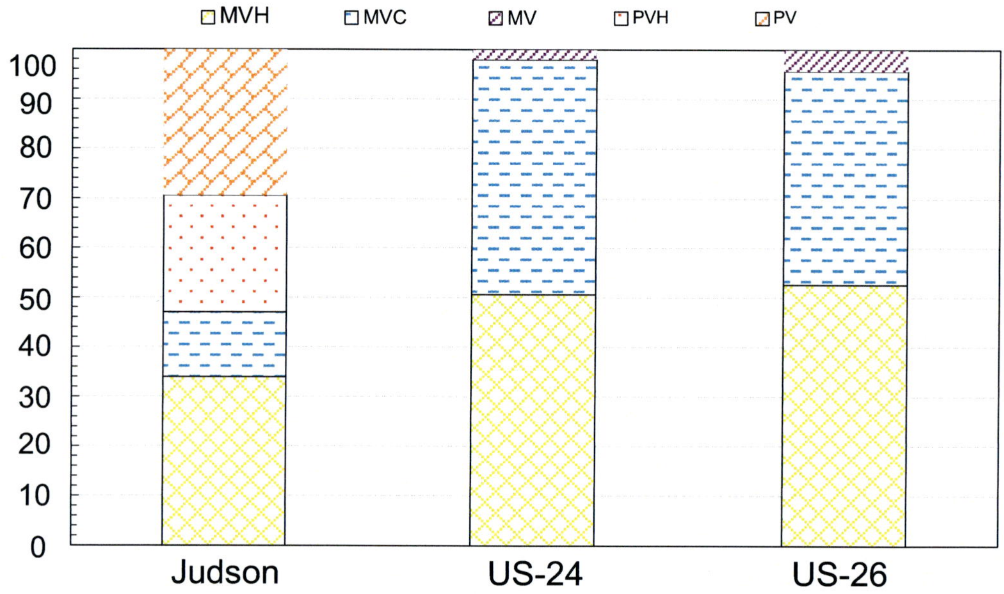

Fig. (5.5). Comparison, of the percentage of operational hours in various modes, between the Judson College building and a US standard building with a cooling set point 24°C (US-24) and 26°C (US-26). [9].

Traditional homes in the Arab world are often noted for their moderate indoor temperatures both in winter and summer. Such buildings, centered on a courtyard and clad with large *mashrabiyat* (lattice-work windows), fountains, thick walls and white marble floors are abundant with passive solar design techniques.

A case study in Damascus, Syria shows how traditional building designs can be integrated into modern buildings to increase energy efficiency. In 2008, a French school in Damascus incorporated several common elements of traditional Arab building design in an effort to create an energy-efficient building. A special focus was placed on ventilation and cooling methods in light of the arid environment. The campus, shown in Fig. (**5.6**), consists of clusters of small, two-story buildings connected by landscaped courtyards that are covered with removable shading. The removable shading is closed above the courtyard during summer days to protect it from the sun, and opened at night for cooling. The opposite process is used for the winter season [10].

Solar chimneys are used to pull the warm air out of classrooms and replace it with cool air. Earth ducts are embedded in the ground floor to pre-cool the air temperature before it enters a room by maximizing contact with the earth which maintains a constant temperature. This particular ventilation strategy finds its origins in traditional Iranian architecture which utilized outlet wind towers and underground chambers where air was cooled by its contact with the earth [10].

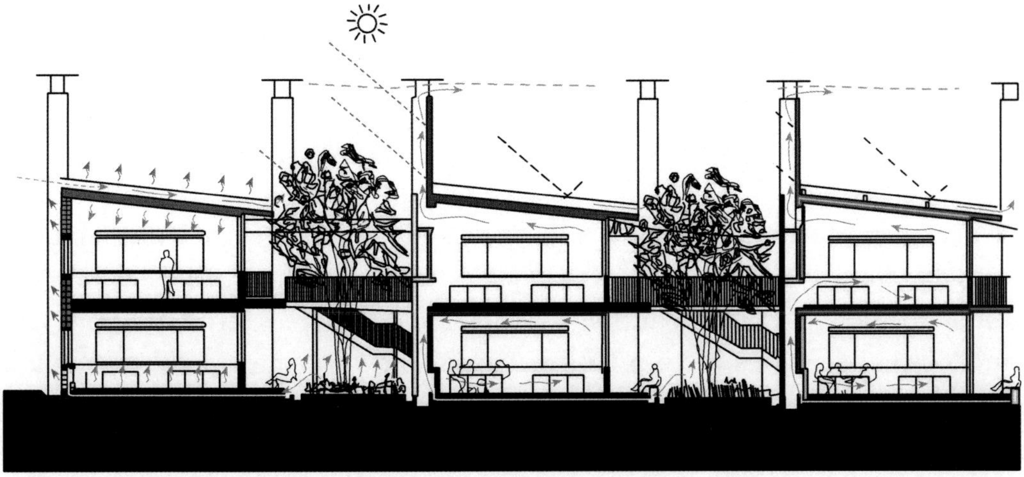

Fig. (5.6). Part section through the Damascus passive solar school [10].

Because of the diurnal swings of desert climates, high thermal mass used in traditional architecture helps by absorbing heat during the day and preventing it from entering the building, and releasing this heat during the cooler nights. The solar chimneys in the Damascus school, shown in Fig. (**5.7**), provide ventilation throughout the building. They are oriented towards the south and are covered with dark sheets to attract solar radiation at the top of the chimney. The heat at the top of the chimney pulls warm air from the room below. Wind creates pressure at the top of the chimney, increasing the movement of air. At night, the thermal mass of the chimney releases heat, continuing the ventilation and cooling process. Fig. (**5.4**) below illustrates the mechanisms of the solar chimney used in this building [10].

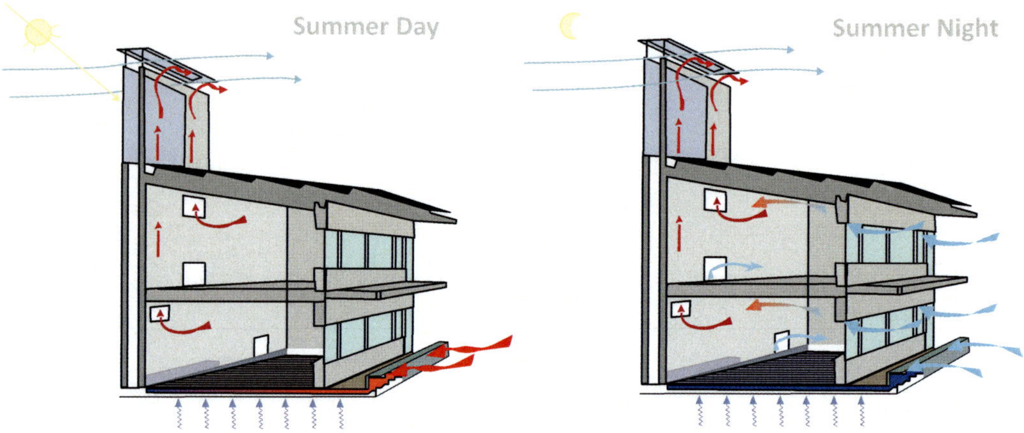

Fig. (5.7). Part section showing the solar chimneys in the Damascus school [10].

This example of the integration of traditional architecture mechanisms, which are essentially passive solar design techniques, shows how old methods can be used in the construction of new buildings to make them more energy efficient. This example is in line with all four pillars of sustainable development, include cultural factors. Encouraging the use of such techniques is economically efficient, environmentally friendly and socially acceptable.

SOLAR THERMAL

Figs. (**5.8** and **5.9**) show that the massive installed capacity of solar water heaters in operation is in China and Europe, while 1.8% of installed capacity is in MENA region. Fig. (**5.10**) shows that most of the MENA region falls in the Sunbelt region, extending within latitude +35(and -35(, with the most favorable conditions for solar energy applications. Ninety percent of the solar radiation in this region comes from direct beam radiation and over 3,000 hours of annual sunshine [11]. These countries can easily depend on solar radiation as a reliable source of energy that can be readily and cheaply exploited by both rural and urban households for a multiple of purposes. Fig. (**5.9**) shows that even though the MENA region geographic location make it a leading candidate for solar energy most of the installations are in only 5 out 21 states.

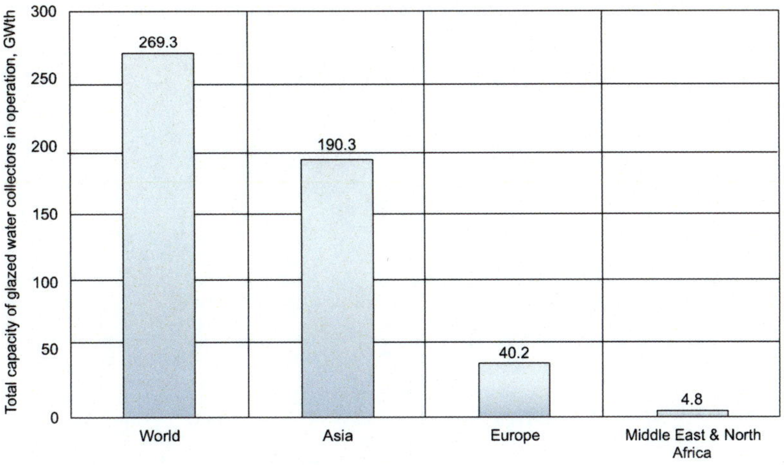

Fig. (5.8). Total installed capacity of glazed water collectors, 2012 [12].

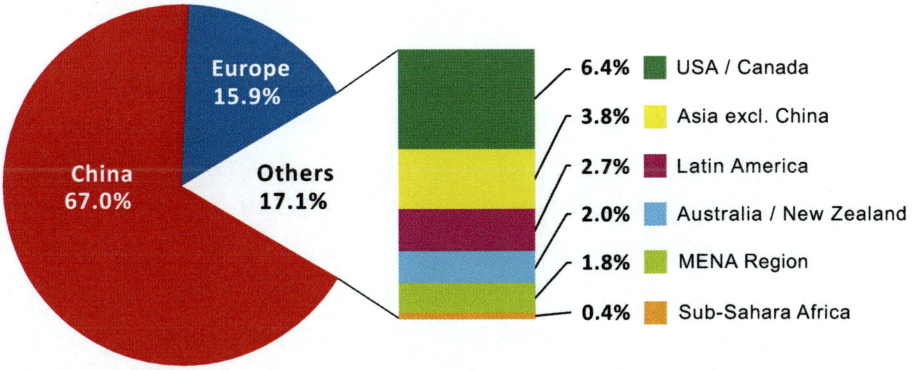

Sub-Sahara Africa:	Mozambique, Namibia, South Africa, Zimbabwe
Asia excluding China:	India, Japan, Korea South, Taiwan, Thailand
Latin America:	Brazil, Chile, Mexico, Uruguay
Europe:	EU 28, Albania, Macedonia, Norway, Switzerland, Russia, Turkey
MENA Region:	Israel, Jorden, Lebanon, Morocoo, Tunisia

Fig. (5.9). Share of total installed capacity of glazed water collectors, 2012 [12].

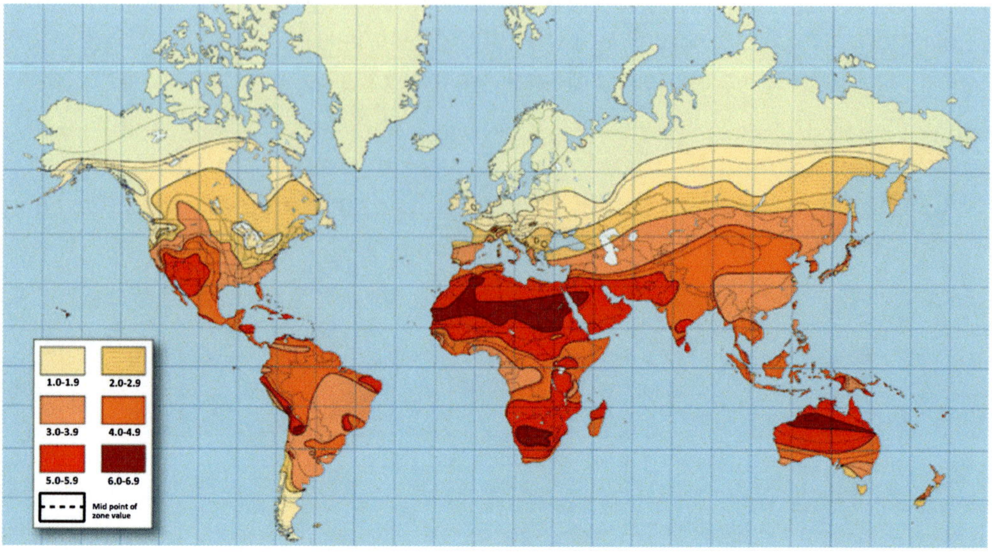

Fig. (5.10). Solar insolation (kWh/m^2/year) on an optimally tilted surface during the worst month of the year [13].

An analysis of the total worldwide installed 269.3 GW_{th} capacity in operation in 2012 by collector type shows that 26.4% of the installed capacity is glazed flat-plate collectors, 64.6% is evacuated tube collectors, 8.4% is unglazed water collectors, and 0.6% is glazed and unglazed air collectors [12].

Flat plate solar-powered water heaters are relatively simple to construct and could prove very useful for a wide range of community types in MENA region. These heaters could be easily made out of local, recycled materials, as proven by the Manshayet Nasser and Darb al-Ahmar communities in Cairo, Egypt. The photographs below, in Fig. (**5.11**), show a locally-made solar water heater in Darb al-Ahmar, Cairo, Egypt. Approximately 24 pilot projects have been constructed in Darb al-Ahmar and Manshayet Nasser by the non-governmental organization (NGO) Solar Cities. In addition, several solar powered technology companies have donated heaters to families in both districts. Locals who use the heaters hail them as successful, and others in the area are open to the use of similar heaters. One of the problems cited by a family which uses a heater to run their *Hammam* (public bath) is that damage to the heater is not conveniently repaired, as their heater was installed by an NGO, with limited funding, rather than a company. For many families in this area, the solar power water heaters either replaced a long process of gathering and burning garbage to heat water, or lowered the long term costs of heating water *via* gas powered heaters.[1]

Fig. (5.11). Solar water heater in Darb al-Ahmar.

Additionally, a UNDP-funded pilot project involving solar water heaters was carried out in El-Menia, Upper Egypt from 1998 to 2000. The project installed 163 water heaters, and gave training sessions on how to use the heaters. Families were required to pay for a portion of the costs, with the goal of instilling a sense of ownership. The heaters installed were flat-panel solar water heaters with a capacity of 150 liters. In the Menia communities in which these heaters were installed, many residents were heating water using agricultural residues, leading to local environmental and health damages. The project was deemed successful in providing families with a clean energy sources, as well as raising awareness about renewable energy and related technologies [14]. As discussed above, such projects have proven successful in underprivileged urban and rural centers in Egypt, and if expanded on a wide scale could contribute more significantly to the energy needs of the country and region.

SOLAR ELECTRICITY AS A RELIABLE SOURCE

Fig. (**5.12**) shows that the EU countries lead the world in PV installations with about 60% of installed PV worldwide capacity, in 2013. On the other extreme comes the MENA region, least in the ranking with only 0.68% of installed PV worldwide capacity.

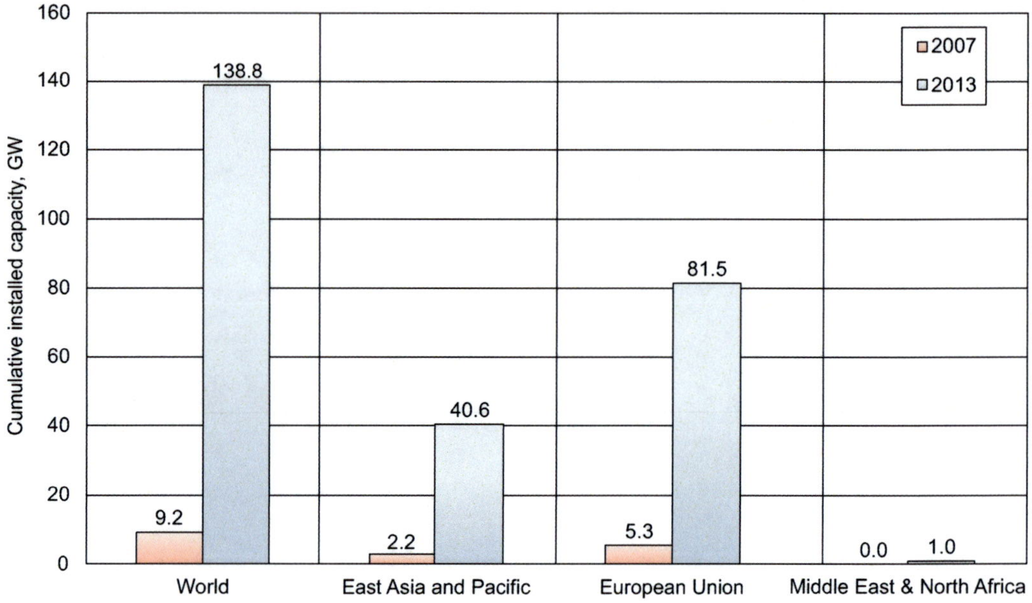

Fig. (5.12). Cumulative installed capacity for PV [15].

Of all solar energy that reaches the earth, the MENA region receives between 22% and 26% of this energy. The annual potential of solar energy per square kilometer received by the MENA region is equivalent to what is generated from 1(2 million barrels of oil [16]. Solar electricity could prove useful not only for the metropolitan areas, but for remote areas where electrical power is not reliable or where electrical grids are erratic or non-existent. This is especially the case for conflict-ridden areas, such as the Gaza Strip or the West Bank where we have seen a particular increase in locally-made solar powered technologies.

In 1999, a UNDP-funded pilot project was carried out in a remote, rural community in the West Bank where a significant number of communities do not have access to electricity which was expensive for those who do have access. Through this project, electricity was generated for the town of Innab Alkabeera using photovoltaic panels and was used to meet daily needs including the provision of power for lights, refrigerators, washing machines, radios, televisions and phones. The panels met the needs of 10 homes as well as a local mosque and school [14]. The project saw heavy involvement from the community, which also worked to spread awareness among other communities out of hope that the project would be repeated elsewhere. With proper maintenance, the system was expected to provide electricity for the town for at least 25 years. Members of the town were given instructions on how best to use the available electricity efficiently as to make the most out of what is available to them [14].

The system used for this pilot project was a solar/wind hybrid with 42 photovoltaic panels as well as a wind generator. Each household installed a meter to monitor individual electricity use and aimed to keep it below a certain amount. Electricity bills were structured as to reward community members for efficient use of energy. The project greatly enhanced the community on several fronts. It allowed the community to refrigerate food and medicine, enhanced education by providing lighting in the local school, and increased availability of information through television and radio. With public lighting, members of the community felt safer and were able to function during evening hours [14]. This project was successful on several fronts and could serve as an example for other remote communities in the Arab world.

SOLAR WATER DESALINATION FOR COASTAL COMMUNITIES

It is imperative that the countries of the Arab world, in addition to conserving water, search for alternate water sources. One-fourth of the world's population does not have adequate access to fresh water. Freshwater sources in the region, which is particularly arid, are already under pressure from demographic growth and economic development. Basic living standards are met at the international water scarcity limit of 1000 m^3/capita/year. Fig. **(5.13)** shows the renewable internal freshwater resources per capita in different parts the world [17]. The figure shows that the annual renewable freshwater resources for the Middle East falls in the "Chronic Water Scarcity" region on the Falkenmark index [18] and is almost 70 m^3/capita/year higher than "Absolute Scarcity" limit of 500 m^3/capita/year described as "a region beyond the 'water barrier' of manageable capability", on the Falkenmark index.

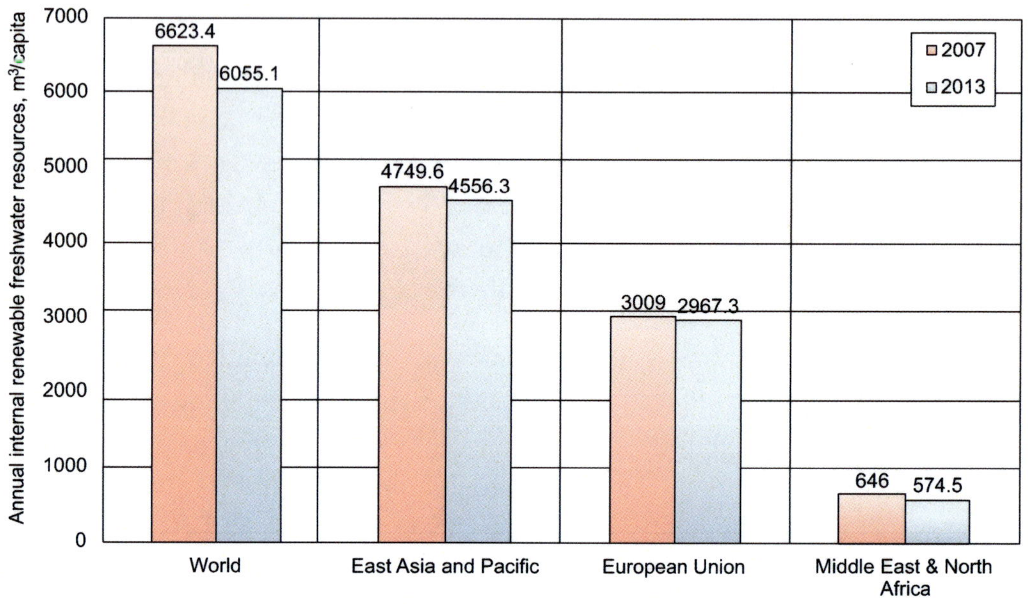

Fig. (5.13). Renewable internal freshwater resources per capita [17].

Fig. (**5.14**) shows the 2013 data for the internal renewable freshwater resources [19], river flows from other countries are not included because of data unreliability. The figure shows that most of the Arab countries are below the lowest limit of the international water scarcity and are expected to fall even further below as their populations continue to grow.

Regional asymmetry and dependency on water availability has been a source of political conflict in the region. High water demand and unsustainable extraction levels exacerbate Israeli-Syrian, Israeli-Palestinian, and Turkish-Syrian tensions [20]. Many countries in the Middle East border one or more inexhaustible water source in the form of an ocean or sea and additionally have abundant solar energy sources. Given this combination, the process of solar desalination, removing salt from water, has the potential to enhance environmental and humanitarian stability in the region. Large-scale, fossil-fuel powered desalination uses a large amount of energy and complex infrastructure, making the process expensive and unsustainable. If better developed, solar desalination could provide a cheaper alternative to traditional desalination processes.

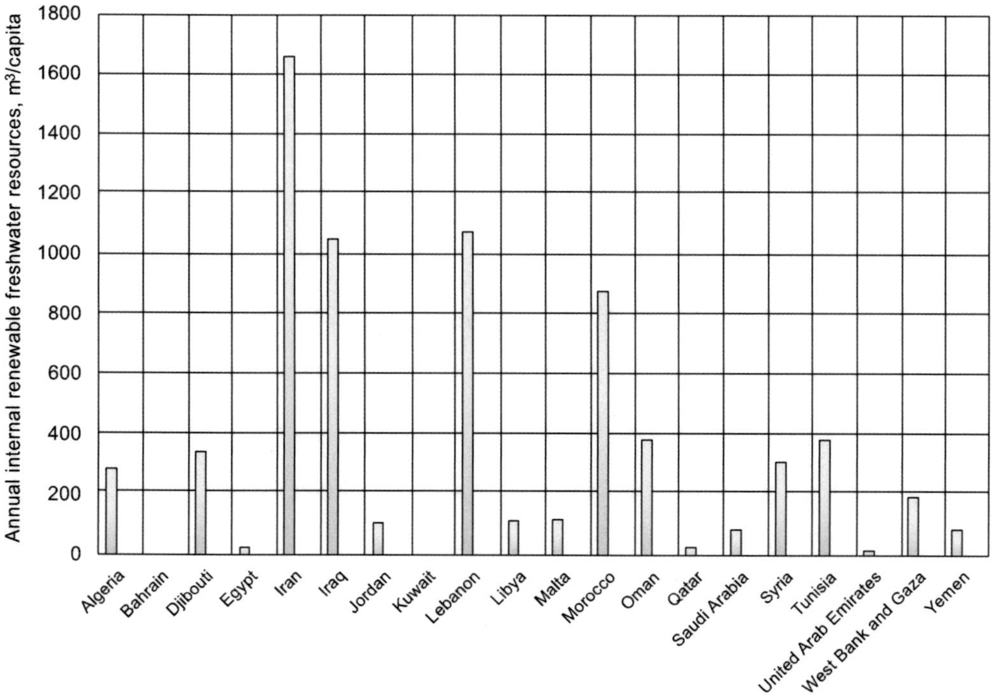

Fig. (5.14). Water availability in MENA region for 2013 [19].

While most large-scale solar desalination technologies remain inefficient in terms of the cost of space and material, many small-scale technologies are considered efficient and would be especially useful for remote and coastal communities. In the Arab world, there are very few examples of solar desalination projects. A pilot project was conducted on the African island of Mauritius which oversaw the construction of small-scale solar stills to desalinate water for 21 families. The units were locally manufactured and used in remote communities. The solar stills used consisted of a reservoir covered with glass, tilted at an angle, allowing the freshwater to drain out into another receptacle. The freshwater derived from the process was approximately half the amount of the original seawater placed into the system, and must still be treated with minerals before drinking. This small-scale method proved successful in this remote Mauritius community, and could very well be adapted in many remote, coastal communities in the Arab world as a step towards conserving crucial natural water resources [14].

SOLAR DRYERS FOR ECONOMIC DEVELOPMENT

Solar dryers can be especially useful for rural areas in the Middle East and beyond. In rural Egypt, the drying of rice and other on blankets in the open air can be seen commonly during crop-picking seasons. A case study was conducted in 2011 on the use of solar dryers for drying agricultural crops in rural Palestine with funding from the United Nations Development Program (UNDP) [14]. A community in Gaza where women traditionally dry crops, herbs and fruits in the sun was selected. As discussed in Chapter 4, the process of drying these herbs in the sun takes away from their freshness and increase risks for contamination from animals, insects or dirt. Simple solar dryers can be locally constructed and result in improved agricultural products and a more efficient drying process. In this particular Gaza community, locally made solar dryers, were installed in the backyards of several households; each dryer was shared by three to five families. The dryers replaced open-air drying and drying by use of gas-powered ovens. The women gathered the crops, fruits and herbs that they had either grown or bought and used the solar dryer to dry them, after which they could keep what was needed for their household and package and sell the remainder. This was especially beneficial for the households involved, as they had produced an enhanced, cleaner product for themselves, saved time for other activities, and increased their incomes [14]. The success of the project in this rural community could provide a useful example for other villages in the region which face similar problems. The solar dryer is simple, inexpensive and could support rural labor if locally manufactured.

SOLAR COOKERS IN CASES OF NEED

Cooking food without causing harm to the environment can be done safely and conveniently using solar cookers. More practically speaking, solar cookers are most important for the portion of the world's population which currently cooks over fires fueled by wood or other materials. In Cairo, residents of slums such as

Manshayet Nasser gather and burn garbage for cooking purposes, consuming a great deal of effort and resources in the process. Additionally, such methods of cooking present a health hazard for an entire community. For this portion of the world's population, solar cookers would be an affordable and much needed change in lifestyle.

Moderate temperatures maintained by solar cookers means that food does not need to be stirred and does not burn, but also entails a longer cooking time. In the solar cooker, food can be easily left to cook unsupervised for many several hours. To allow for solar cooking while for example at work sites or while pursuing an activity that is outdoor, many designs of the cooker were made to be portable. The basic Solar Box Cooker, for example, can simply be constructed of cardboard or other readily available materials at very little expense [21].

A very low-tech approach to solar cooking is digging a pit in the ground that is shallow while insulating the bottom using leaves or dried grass. Then the food or water is placed in a dark vessel with a clear cover such as glass on the top. Applying the same solar principles on the high-tech end of the scale, a solar cooker can be easily integrated into the south-side of a contemporary kitchen using standard building and insulating materials and high performance glazing [21].

Due to the wide availability and in expensive cost of cardboard, cardboard solar box cookers could be suitable for various cultures. However, when compared to other materials, vulnerability to moisture damage and lack of durability are major drawbacks cardboard. Aesthetics are usually valuable. This is because the whole solar cooking concept may be declined in certain cultures that have the norm as a rounded form since the solar cooking box is a square. In addition to that, some certain social strata that would reject the cardboard material as it is a "cheap material" for their usage [21].

The appropriate solar cooker for a family or individual depends on several factors, including financial abilities and location (urban or rural). For wealthier families, particularly those which reside in rural settings, it may be possible to integrate a solar cooker in the architecture of the house kitchen, as mentioned above. For wealthier families, both in urban and rural settings, an easily transportable box cooker could be used for picnicking purposes [21].

In terms of social groups which burn wood or other materials to heat food or pasteurize water, solar cookers are a necessity rather than luxury. Although both the parabolic and panel cookers are efficient, they are more difficult to construct and less portable than the solar box cooker. Solar box cookers are highly efficient; one which accounts for sufficient interior space, durability and transportability would be ideal for such social groups [21].

Solar cookers have been used in the Gaza Strip as a means of survival, as the blockade imposed on the territory often prevents locals from using traditional energy sources. In 2008, a family in the town of Deir Elbalah used local materials and a dysfunctional gas oven to construct a solar cooker. The family used the structural portion of the oven as well as mud, bricks and straw to increase insulation. They additionally used two sheets of glass, one for the top of the oven and the other for a side as well as a metal sheet for the bottom of the oven. The solar oven was successful, and the same design was imitated by approximately 45 families in the town. These solar ovens are not only environmentally friendly, but provide a basic need for these families at little cost [22]. This example took into account local needs and conditions, and can be applied to other areas in the region facing a similar lack of access to energy. Remote and itinerant communities could especially benefit from this example.

ZERO ENERGY COMMUNITIES

By combining the technologies discussed in this book as well as other renewable energy technologies, an entire community that relies minimally on fossil fuel energy can be created. Several communities have been launched with the aim of minimizing the need for fossil fuel energy sources and maximizing the use of renewable energy where possible. These communities have shown the vast potential for the implementation of alternative energy technologies.

One such community, nearly a decade old now, is located in Beddington, England. The project is referred to as the Beddington Zero (Fossil) Energy Development, or BedZED, (Fig. **5.15**). The settlement was designed from scratch for the purpose of exploring the potential of creating a more sustainable lifestyle that simultaneously meets modern standards of comfort. The overall goal of the project was to reduce carbon emissions through both reducing energy consumption and using alternate energy sources. Specifically, the project set out to reduce electricity and water consumption as well as energy consumption needed for transportation and indoor heating and cooling [23].

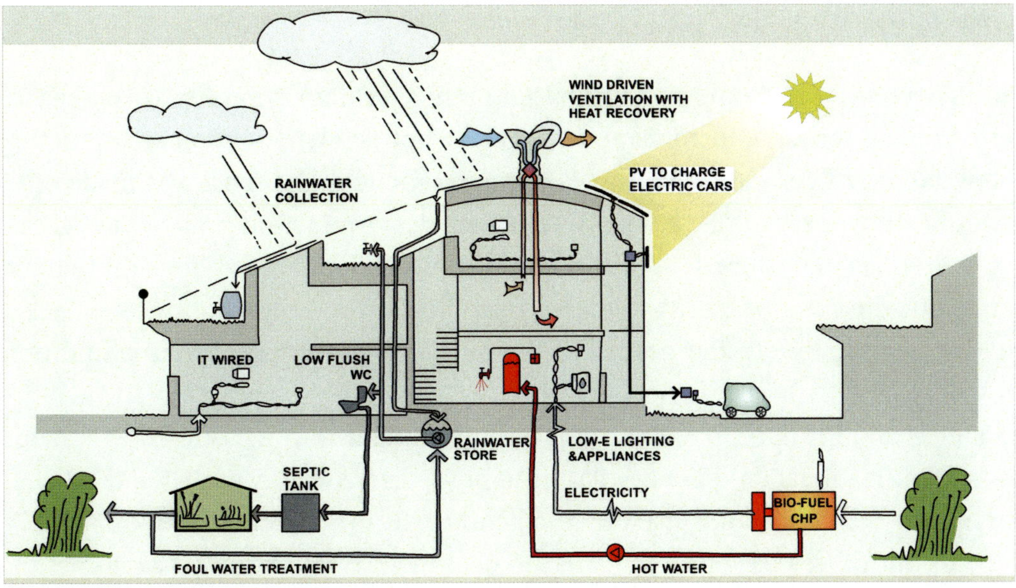

Fig. (5.15). BedZED zero energy system [23].

As it was important to the architects of BedZED to minimize the need for energy, buildings were constructed according to several of the principles discussed in this book's sections on solar passive buildings. Specifically, workspaces featured long, north-facing, windows to reduce the need for artificial lighting during the daytime. The level of insulation is particularly high in BedZED buildings, to reduce the need for heating. To further reduce heat loss, buildings feature triple-glazed windows, and windows and doors are well-sealed. The materials used for these buildings were recycled where possible, to reduce the implicit need for energy consumption or habitat destruction to secure new materials. Also where possible, the materials were obtained from nearby locations to reduce energy costs associated with transportation [23].

When it came to the design of the community, architects aimed to create a mix of residential, work and community spaces within a specific area. This would greatly reduce the need for transportation within a given area and enhance the lifestyles of residents. Furthermore, architects took advantage of the space provided to incorporate plant and wildlife where possible; for example, rooftop gardens can be found throughout the community. As mentioned previously, plants are crucial to the reduction of carbon dioxide in the atmosphere due to their consumption of the gas for the photosynthesis process [24].

In the selection of technologies to be used in the community, utility was measured against cost in order to determine whether or not implementation was practical. Where the use of certain technologies was not practical, the most was made out of resources. For example, the passive solar design of buildings was greatly emphasized as it reduced or altogether eliminated the need for lighting and heating. In the case of heating, the project aimed for "zeroheating"—a lack of need for heating due to the design of the building along with cooking and other activities that created heat within the house. In addition to building and window orientation and insulation, the houses retain heated air through wind cowls, which rely on similar mechanisms to solar chimneys discussed in the Chapter 4. Less than half of the households use electric heating during the coldest two months of the year, but otherwise buildings are almost entirely reliant on passive methods [23].

PV panels were installed on rooftops of homes; the main purpose of these panels was to energize electric cars. The cars, however, were not popular among residents and only two electric vehicles are used on site. The PV panels were thus used for basic building electricity needs. The success of the panels was not measured precisely, but it was estimated that PV panels contributed to 20% of the community's energy needs. It was relevant to note when considering this estimate that, England was not a prime recipient of solar energy. In order to reduce energy needs to the greatest extent possible, the most energy-efficient appliances available for practical usage were selected for us in the BedZED community. Additionally, meters tracking electricity consumption were installed in households to make residents more aware, and potentially more responsible, users. Due to these methods, in addition to passive solar designs discussed, electricity consumption in BedZED was 45% less than average consumption [25].

BedZED also focused on more sustainable lifestyles in water, transportation and waste. In addition to using technologies which consume water more efficiently, awareness was raised and rain water was recycled where possible to reduce irresponsible consumption. In the area of transportation, plans to encourage the use of the electric car were not successful. However, the design of the community itself lessened the need for transportation for many, and more efficient methods, such as carpooling, were encouraged. Minimizing waste and recycling when possible were also important aspects to the sustainability of the community.

BedZED is an important illustration of the potential holistic use of renewable energy technologies. The community is not without its flaws; however, more practical renewable energy technologies are being discovered by the year and it is likely that a community built at the present time or in the future could be even more sustainable than BedZED. The community's location, in an area that is far from the most ideal in terms of solar and wind energy availability, further suggests the even greater potential for such a community in other regions of the world.

The project of Masdar city in Abu Dhabi is another example which illustrates sustainable energy, (Fig. **5.16**). This city was designed so that it would be the first city in the world that is carbon neutral. The city is 6 sq. km and utilized several renewable energy and eco-city plans [26]. The buildings used 54% less water and 56% less energy. This was achieved by using the latest efficient technologies and passive building methods [27].

DISCUSSION

This chapter has shown, through the presentation of case studies from different geographic locations, that a technology must take into consideration factors beyond technological performance in order to be truly sustainable. The role of a technology in relation to the local society, economy and culture is as crucial as the successful performance of that technology itself. As we have seen, necessity has often led to the creative use of these technologies in the region. We should not wait until resources are unavailable or environmental or human health is harmed to a degree that makes implementation of these technologies an absolute necessity. By starting to use these technologies now, at least on a small scale, we can begin to combat the detrimental effects of other energy sources and improve upon existent models through practical experience. The implementation process, however, is not without challenges. These will be discussed Chapter 7.

Fig. (5.16). Ground views of Masdar City showing the sand-colored Arabic architecture residential buildings [27].

QUESTIONS

1. What are the four pillars of sustainable development?
2. How does the passive solar building example of the Damascus school account for all four factors of sustainable development?
3. What impact did solar dryers have on society and the economy in the Gaza community discussed?
4. How did the use of solar cookers and solar electricity enhance life in the communities discussed?
5. In what types of communities can the examples discussed be emulated?

6. How do the case studies discussed benefit environment, health, society and the economy?

FOOT NOTE

[1] Interview with Darb al-Ahmar Community Members. November 2010.

REFERENCES

[1] World Commission on Environment and Development, "Report of the World Commission on Environment and Development: Our Common Future (The Brundtland Report)", *UN documents, Available: http://www.un-documents.net/wced-ocf.htm,* 1987. [Online]

[2] C. Barras, Why Sustainable Power is Unsustainable, "New Scientist", *Available: http://www.newscientist.com/article/dn16550-why-sustainable-power-is-unsustainable.html#.VX1hyvmqpHw.*Feb-2009 [Online]

[3] B. Bordass, R. Cohen, M. Standeven, and A. Leaman, "Assessing Building Performance in Use 3: Energy Performance of the Probe Buildings", *Build. Res. Inform.,* vol. 29, no. 2, pp. 114-128, 2001. [http://dx.doi.org/10.1080/09613210010008036]

[4] Y. Ji, K.J. Lomas, and M.J. Cook, "Hybrid Ventilation for Low Energy Building Design in South China", *Build. Environ.,* vol. 44, no. 11, pp. 2245-2255, 2009. [http://dx.doi.org/10.1016/j.buildenv.2009.02.015]

[5] K.J. Lomas, M.J. Cook, and D. Fiala, "Low Energy Architecture for a Severe US Climate: Design and Evaluation of a Hybrid Ventilation Strategy", *Energy Build.,* vol. 39, no. 1, pp. 32-44, 2007. [http://dx.doi.org/10.1016/j.enbuild.2006.03.032]

[6] B. Krausse, M. Cook, and K. Lomas, "Environmental Performance of a Naturally Ventilated City Centre Library", *Energy Build.,* vol. 39, no. 7, pp. 792-801, 2007. [http://dx.doi.org/10.1016/j.enbuild.2007.02.010]

[7] C.A. Short, K.J. Lomas, and A. Woods, "Design Strategy for Low-Energy Ventilation and Cooling within an Urban Heat Island", *Build. Res. Inform.,* vol. 32, no. 3, pp. 187-206, 2004. [http://dx.doi.org/10.1080/0961321041000l679875]

[8] M.W. Simons, and B.J. Maloney, "Comfort and Acoustic Monitoring in a Large Naturally Ventilated Technically Advanced Building", *Int. J. Vent.,* vol. 2, no. 1, pp. 1-13, 2003.

[9] C.A. Short, and K.J. Lomas, "Exploiting a Hybrid Environmental Design Strategy in a US Continental Climate", *Build. Res. Inform.,* vol. 35, no. 2, pp. 119-143, 2007. [http://dx.doi.org/10.1080/09613210600852789]

[10] K. Elgendy, "A Damascus School Revives Traditional Cooling Techniques", In: *Carboun.* 2010. Available at: https://www.google.com.pk/webhp?sourceid=chrome-instant&ion=1&espv=2&ie=UT--8#q=A+Damascus+School+Revives+Traditional+Cooling+Techniques

[11] S. Romero-Hernandez, B. Rodríguez-Granada, O. Romero-Hernandez, and D. Wood, *Solar Energy Potential in Mexico's Northern Border States..* Woodrow Wilson International Center for Scholars, 2012.

[12] F. Mauthner, and W. Weiss, "Solar Heat Worldwide: Markets and Contribution to the Energy Supply 2012", In: *International Energy Agency (IEA) Solar Heating and Cooling Programme, Gleisdorf.* Austria, 2014.

[13] The Ugly Side of Solar Panels, "LOW-TECH MAGAZINE", *Available: http://www.lowtechmagazine.com/2008/03/the-ugly-side-o.html.*Accessed: 18-Jun-2015 [Online]

[14] E. Ebrahimian, ""Community Action to Address Climate Change: Case Studies Linking Sustainable Energy Use with Improved Livelihoods," Global Environment Facility (GEF)", In: *Global Environment Facility (GEF) Small Grants Programme,* . United Nations Development Programme, 2003.

[15] "Global Market Outlook for Photovoltaics 2014-2018", *European Photovoltaic Industry Association (EPIA),* 2012.

[16] J. Asnani, "IFC Knowledge Series in MENA, The Potential of Renewable Energy in MENA", *International Finance Corporation (IFC).*

[17] ""Renewable Internal Freshwater Resources per Capita", *Worldbank. Available: http://data.worldbank.org/indicator/ER.H2O.INTR.PC/countries/1W-ZQ-EU-Z4-A9?display=default* .Accessed: 14-Jun-2015

[18] P. Lawrence, J. Meigh, and C.A. Sullivan, *The Water Poverty Index: An International Comparison..* Centre for Economic Research, Keele University: Staffordshire, UK, 2003.

[19] *World Development Indicators 2015,* World Bank Publications, 2009.

[20] A.M. Michael Mason, *Renewable Energy in the Middle East: Enhancing Security through Regional Cooperation..* Springer Science & Business Media, 2009.
[http://dx.doi.org/10.1007/978-1-4020-9892-5]

[21] M. Aalfs, "Principles of Solar Box Cooker Design", *UN documents, Available: http://solarcooking.wikia.com/wiki/Principles_of_Solar_Box_Cooker_Design,* 2010.Accessed: 14-Ju--2015 [Online]

[22] R. Bergstein, "Solar Cooking Ovens: Another Eco-Solution from the Gaza Strip", *Green Prophet, Available: http://www.greenprophet.com/2008/11/solar-ovens-gaza-strip/,* 2008.Accessed: 14-Ju--2015 [Online]

[23] C. Twinn, "BedZED", *Arup J.,* vol. 38, no. 1, pp. 10-16, 2003.

[24] *"What Makes an Eco-Town?" BioRegional Development Group and the Commission for Architecture and the Built Environment..* CABE: UK, 2008.

[25] J. Hodge, and J. Haltrecht, *BedZED Seven Years on: The Impact of the UK's Best Known Eco-village and Its Residents..* BioRegional Development Group: London, UK, 2010.

[26] "Sustainable Buildings and Cities Around the World: Masdar City", *Available: http://robswatsonadventure.blogspot.com/2011/03/masdar-city.html.*Accessed: 20-Jun-2015 [Online]

[27] P. Patel, and S. Griffiths, "Masdar City Showcases Sustainability", *Mater. Res. Soc. Bull.,* vol. 38, no. 06, pp. 450-451, 2013.
[http://dx.doi.org/10.1557/mrs.2013.140]

CHAPTER 6

Natural Lighting in Buildings

Khaled Nassar*

The American University in Cairo, Department of Construction Engineering, Cairo, Egypt

Abstract: Natural lighting is an essential element in successful design in order to decrease the electrical lighting energy consumption while ensuring visual comfort such as the reduction of glare. This chapter presents the topic of daylighting as it pertains to the urban scale. The chapter first provides an overview of the several benefits of daylighting in various applications. Some fundamentals about daylighting are then presented. This is followed by the presentation of the concept of solar envelopes as an important aspect of daylighting in the urban scale. Then different studies pertaining to daylighting at the urban scale are presented.

Keywords: Climatic daylighting metrics, Daylighting, Dynamic daylighting metrics, Energy, Fresnel lenses, Light redirection, Light wells, Lighting, Radiance, Simulation, Sky types, Solar tubes, Urban design, Urban sprawl, Window treatment.

INTRODUCTION

Daylighting is the process of trying to provide natural sun light to internal and external spaces during the day through the design of openings, surfaces, fenestrations, facades, building elements and other active elements. The main goal is to increase the amount of natural light while decreasing electrical lighting and ensuring visual comfort such as the reduction of glare. This chapter presents the topic of daylighting as it pertains to the urban scale. The chapter first provides an overview of the several benefits of daylighting in various applications. Some

* **Corresponding author Khaled Nassar:** Department of Construction Engineering, School of Sciences and Engineering, The American University in Cairo, AUC Avenue P.O. Box 74, New Cairo 11835, Egypt; Email: knassar@aucegypt.edu.

Salah El-Haggar, Sarah Mousa & Mohamed El-Morsi (Eds.)
All rights reserved-© 2016 Bentham Science Publishers

fundamentals about daylighting are then presented. This is followed by the presentation of the concept of solar envelopes as an important aspect of daylighting in the urban scale. Then different studies pertaining to daylighting at the urban scale are presented.

BENEFITS OF NATURAL LIGHTING

Providing adequate natural light in building spaces and also outdoors is a very important aspect for several reasons. Firstly, providing natural sunlight can reduce the amount of electrical lighting required to provide the desired level of illumination in and around buildings. This in turn can reduce the amount of energy required for electrical lighting and thus reduce the amounts of harmful emissions resulting from electricity generation. Consider for example that the United States consumes about 412 billion kilowatt-hours (kWh) of electricity for commercial and residential lighting. This means that 15% of the residential and commercial electricity consumption was attributed to lighting [1].

In particular, institutional and commercial buildings, as well as street lighting used more lighting energy in 2014 than the residential sector totaling about 262 billion kWh (19% of the total electric energy consumed). Residential lighting consumption on the other hand was about 150 billion kWh (14% of total electric energy consumed). The impact of reducing the amount of electrical energy assigned to electrical lighting by enhancing the use of natural lighting becomes clear.

Secondly the health benefits of daylighting have long been known. In principle, light in general and daylight in particular orchestrates the circadian response and influences the metabolism and the endocrine and hormone systems as well as the production of melatonin, dopamine, and serotonin [2]. Daylight additionally is partly responsible for Vitamin D in our bodies by way of photosynthesis through our skins. This is a very important factor in bone diseases and Multiple sclerosis. It is also well known that light has an impact of what is known as Seasonal Affective Disorder (SAD) as well as depression, stress and anxiety.

In addition to the above benefits, light also affects the quality of life and aesthetics of the building and urban spaces. Therefore daylighting legislation in different countries have been implemented to assure adequate daylight. Most countries have in their building ordinances requirements for solar zoning legislation, which basically sets certain limits on setbacks, street widths, and building heights to ensure that building spaces have adequate access to sky and/or the sun. In the English law for example there are provisions for what is known as "right to light" which is a form of easement which guarantees that buildings with windows maintain the same level of access to daylight with any addition of modifications to adjacent buildings. In addition most building codes will set requirements for the presence of windows as well specifying their sizes. Additionally, sustainability certification schemes such as the US green building rating system LEED, will award buildings which achieve certain levels of daylight.

For specific building types, daylight has special importance. For example, it has been shown [3, 4] in schools that children score better on tests in naturally lit classroom. In addition to the fact that natural light has been proven to improve standardized test scores, eliminate common distractions and improve health and growth, it also has been shown to reduce late-day energy loss. In the workplace it has shown to increase productivity in has also been proven to increase sales [5]. Daylight has also been shown to help in the healing process in hospitals [6].

DAYLIGHTING FUNDAMENTALS

This section presents important concepts pertaining to daylighting design. The starting point to any daylighting design is understanding the source of the natural light. Natural light comes either from the sky dome or the direct sun beam. In order to simulate and design with daylight, a number of models of the sky and sun have been developed. The most basic sky model is what is known as the Uniform Luminance Model which describes a sky of constant brightness. The main purpose of this model is to represent a heavily overcast sky, where no sun can be seen. However, it has long been known that the brightness of an overcast sky changes gradually from being darker at the horizon and brighter at the zenith.

Therefore, the standard model for modeling overcast sky is the CIE (Commission internationale de l'éclairage or the International Commission on Illumination) Standard Overcast Sky, which was originally known as the Moon and Spencer

Sky. This model better approximates the luminance distribution for overcast skies by providing a gradual change in luminance from the zenith to the horizon where the luminance of the horizon is a third of that at the zenith.

$$L_\theta = \frac{1}{3} L_Z (1 + 2 \sin \theta) \qquad \textbf{(6-1)}$$

Where, L_Z is the sky luminance at the zenith ($\theta = 90°$). Under clear skies, a more complex CIE developed a more complex model to describe the brightness distribution for clear skies including that resulting from the sun. The model accounts for three factors; the bright circumsolar region around the sun, the brightening of the sky near the horizon as well as the dark blue patch in the same azimuth plane as the sun, but at 90 degrees altitude angle from it. The model therefore takes as inputs the position of the sun to determine the luminance distribution of each patch of the sky. Both the CIE overcast standard and the CIE clear sky are normalized to zenith luminance.

In addition to the above three models there are several other sky types. One of the most commonly used is the Perez sky. This sky model accounts for the luminance from the circumsolar region and the light from backscattering, in addition to the three factors of the CIE sky model. The Perez model has been found to be slightly more accurate if the factors above are modeled accurately.

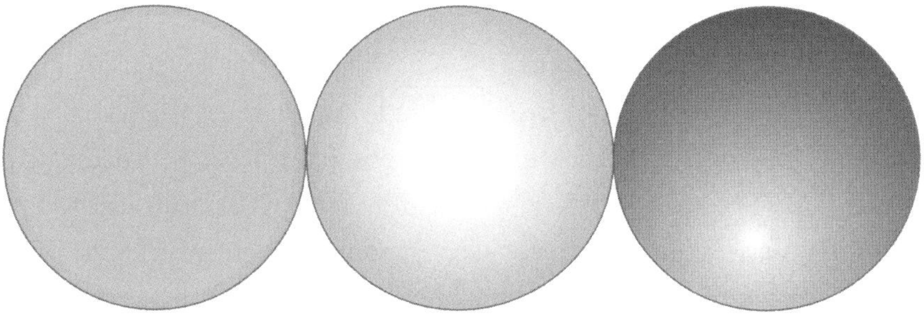

Fig. (6.1). Sky types.

Given an accurate model of the sky, the next issue in daylighting design is to determine a method to simulate light propagation. Here, several methods have been used including graphical methods, point-by-point methods as well as extensive numerical simulations. Software for lighting simulation such as Dialux (and Relux) and the more advanced RADIANCE have been used now for many years in the industry. While Dailux and Relux can perform accurate and fast lighting simulation, RADIANCE specifically is capable of annual hourly simulation using a weather file as well as being capable of modeling blinds/shades operation based on sun penetration and other complex fenestration systems (CFS).

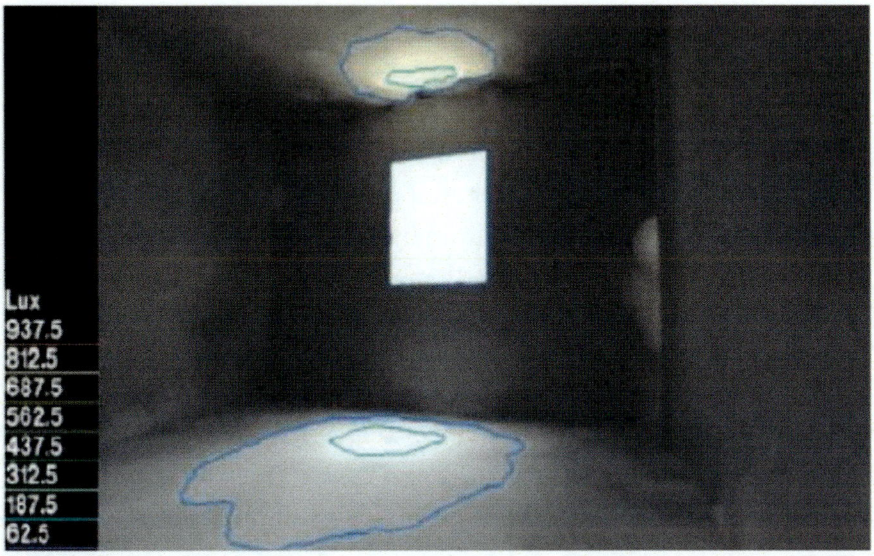

Fig. (6.2). RADIANCE simulation.

Using these software one would be able to determine the illuminance values at various points in the space. Square grid cells are more preferable for illuminance calculations but the aspect ratio of the grid units should be kept between 0.5 and 2 (EN 12193:2007 and EN 12464-2:2007). The limit on the grid size is set to a maximum of:

$$p = 0.2\ x\ 5 \log_{10} d \qquad \textbf{(6-2)}$$

Using these software designers are able to test different design options and examine the results. Metrics to assess these results have been developed. Daylight metrics are essential to determine how to optimize benefits of daylighting while minimizing energy use. Daylight factor has been the metric used since the 1900s. The daylight factor is basically the ratio of illuminance inside the space to the horizontal illuminance outside under the CIE overcast sky. As such, the daylight factor neglects the illuminance from the sun and is therefore not a very good metric to use especially under clear sky conditions. As such the daylight factor is insensitive to orientation and climate of the location. It also does not account for any specific design conditions such as the use of shades or blinds.

Therefore a number of "climate-based" or dynamic measures have been proposed including: Continuous Daylight Autonomy, Daylight Autonomy, Spatial Daylight Autonomy, Zonal Daylight Autonomy, Temporal Daylight Autonomy, Useful Daylight Illuminance, Direct Sunlight Hours, Daylight Uniformity, Maximum Daylight Autonomy, Daylight Saturation Percentage, Annual Sun Exposure and, Annual Light Exposure. Two of those measures in particular were found to be correlate best with occupant assessments in a recent study [7], namely the Spatial Daylight Autonomy (SDA300/50%) and Annual Sunlight Exposure (ASE 1000, 250h).

The Spatial Daylight Autonomy (SDA300/50%) is a measure of daylight illuminance sufficiency for a given area and it represents the percentage of the space floor area that surpasses 300 lux, for 50% of the hours during the year from 8:00 am to 6:00 pm. The "Nominally acceptable" threshold is 55% of analysis area, while the "Preferred" threshold is 75% of analysis area.

The Annual Sunlight Exposure on the other hand evaluates the discomfort from direct sunlight. This measures is the percent of space area that surpasses 1000 lux for more than 250 hours per year. The "Nominally acceptable" threshold is less than 7% of analysis area while the "Preferred" threshold is less than 3% of analysis area.

Therefore north oriented spaces with carefully designed overhangs and higher window head height will perform better in terms of the SDA and ASE. Additionally, spaces that receive less sun penetration, but have deep daylight penetration, and ones that incorporate light redirecting technologies are usually going to receive better SDA and ASE scores. These metrics also awards spaces that offer multi-directional daylight such as having windows on more than one wall and skylights in the ceilings.

On the other hand, the SDA and ASE will be low for spaces without window treatments such as overhangs, blinds, *etc*… which admit too much direct sun. These metrics will also penalizes deep spaces and dark spaces or windows with dark treatments and small windows.

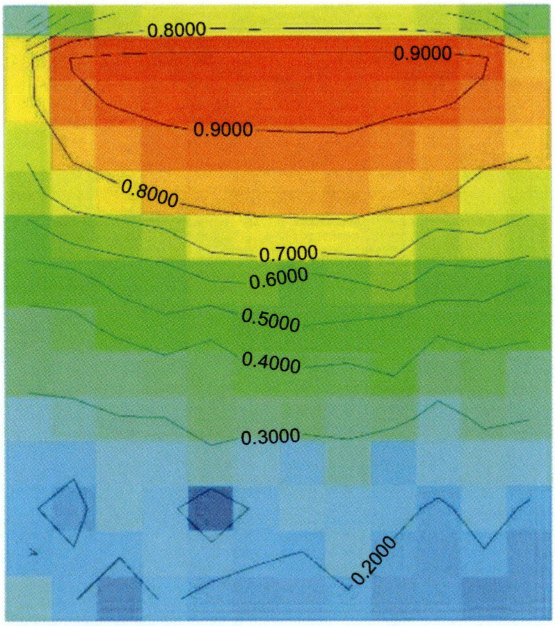

Fig. (6.3). Sample dynamic/climate-based measures.

SOLAR ENVELOPES

Ensuring that buildings and public spaces in an urban scale have access to adequate daylight and sun exposure is an important part of urban planning. One approach to achieving this goal is to use solar envelopes to guide the form of buildings. The solar envelope is essentially an imaginary envelop or shrink wrap for the building which assures solar access for any building within that envelop [8]. The first description of the solar envelop was by Knowles and Berry at the University of Southern California (USC) [9]. The solar envelope is basically guided by the sun's motion. Buildings and structures within this envelop will be provided with solar access and will also not overshadow neighboring buildings during certain times and dates.

The solar envelops dictates the three dimensional form of the buildings with the goal of avoiding shadows during undesired times. This is accomplished by setting a shadow fence along neighboring property lines as well as cut-off times (*i.e.* times during which no overshadowing is allowed) and determining the solar envelop accordingly [10].

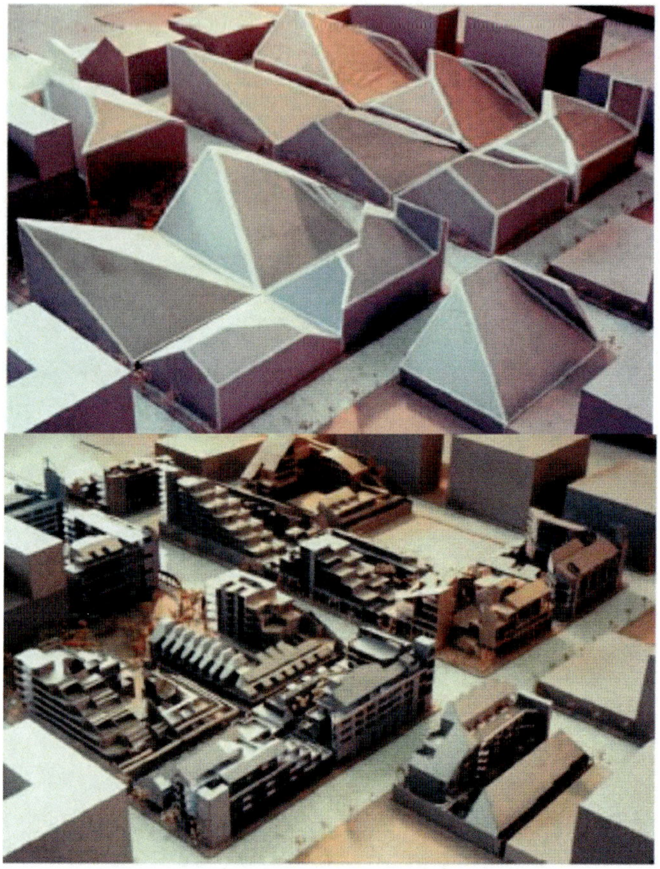

Fig. (6.4). Solar envelopes and the buildings that fit these envelopes [11].

LIGHT AT THE URBAN SCALE

Several techniques are used to provide light at the urban scale, including light-wells, light pockets, mirrors and light redirecting devices. Light-wells are an essential part of the multistory buildings in several parts around the world. In large buildings where the perimeter is small in relation to the area of the building and where deep floor plates are necessary, the internal spaces may not be sufficiently lit. In addition to using inner court yards for lighting livable space, light-wells are often used to provide light and ventilation for bathrooms and kitchens. Light-wells become essential when the building regulations do not allow for mechanically or artificially lit kitchens and bathrooms.

Generally speaking most design criteria and code requirements only focus on the area and the dimensions of the light-wells, setting a minimum area as well as minimum dimensions. This therefore neglects other aspects of the light-well design that affect the performance such as the windows area, the aspect ratio of the light-well, the reflectivity of the surfaces.

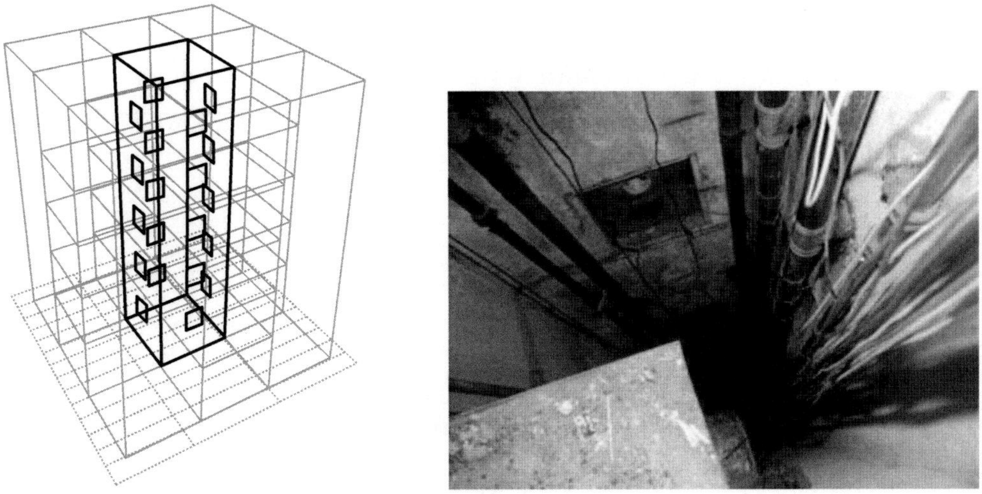

Fig. (6.5). Lightwells.

Previous research on the topic has demonstrated that light-wells with larger aspect ratios tend to become less efficient that those which are closer to a square. Obviously, the larger the light well the better the performance. In one study, three-story building with a light-well was studied [12]. The inner surfaces of the light-well were matt white painted. Light-wells can also improve ventilation as demonstrated by other studies [13, 14].

Although the International Building Code (IBC) specifies that all habitable spaces to have natural or artificial light, in some jurisdictions, natural light is required for all spaces. Artificial light is to be provided at a level of 107 lux over the area measured at about 70 centimeters above the floor level. There is no specific light-quantity requirement set for natural light. This means that in theory, there could be a building that is occupied only during daylight hours that does not have any artificial lighting but instead gets all its light naturally. Note also that the code allows interior rooms to borrow light from adjacent rooms if at least half of the wall between them is open and the opening must be at least 10% of the space floor area of the interior rooms or 2.3 squared-meters.

DISCUSSION

Fig. (**6.6**) shows the effect of changing the orientation of the building on the illuminance of the lightwell in the fourth and third floors. It was found that the illuminance was maximum when the north of the model was oriented at 0 degrees and 90 degrees and it was minimum at 45 and 135 degrees respectively. Also it was found that this effect is very obvious in the fourth floor, however in the four lower floors the effect was nearly not existing.

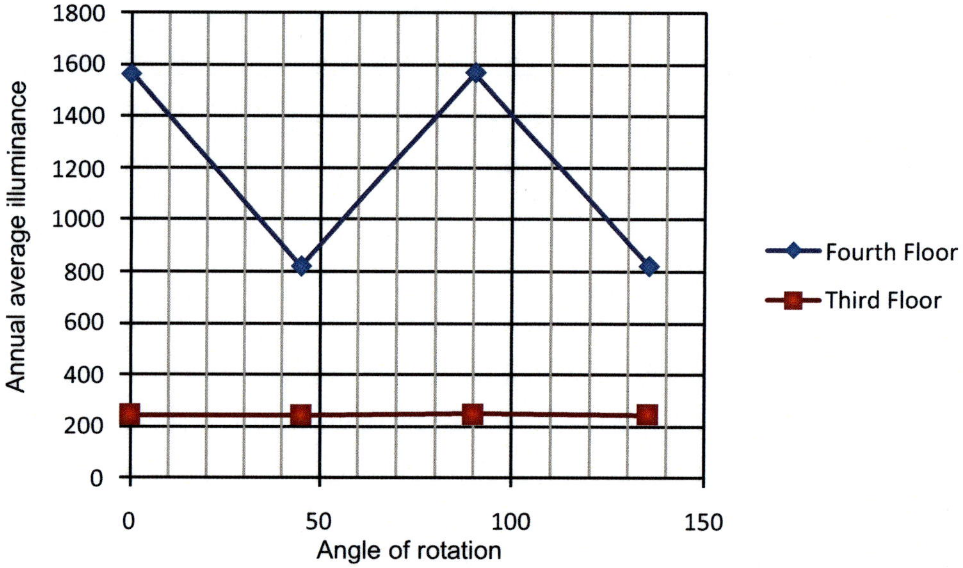

Fig. (6.6). A graph showing the effect of changing the orientation of the building on the luminous flux in the fourth and third floors [15].

Changing the lightwell area was slightly affecting the illuminance of the building. Fig. (**6.7**) shows the effect of changing the lightwell length on the illuminance of the building in the fourth, third and second floors when sweeping the lightwell length from 4 to 7 meters at 0.5 meters step and the lightwell width is fixed at 4 meters. As it is clear from the graph, the luminous flux is increasing at first with increasing the area; however it is decreasing suddenly when the lightwell length is 6 meters in the fourth and third floors. The second floor shows a different behavior as the luminous flux didn't decrease when the lightwell length reached 6 meters. In general, changing the area had a tiny effect on changing the illuminance of the building which is not the case with the effect of the orientation of the building.

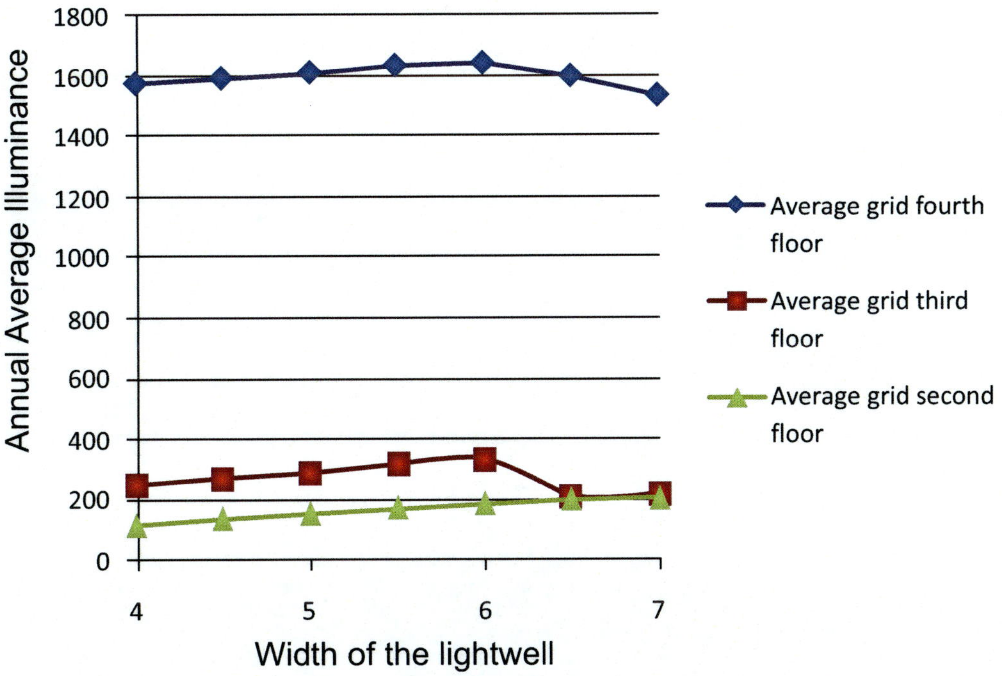

Fig. (6.7). A graph showing the effect of changing the orientation of the building on the luminous flux in the fourth and third floors.

Light pockets are indentation in building facades that also allow more spaces in the building to gain visual access to the outside. The effect of the dimensions of the light pocket was investigated and the resulting effect of pocket width and depth on the average illuminance inside the adjacent rooms is shown in Fig. (**6.9**).

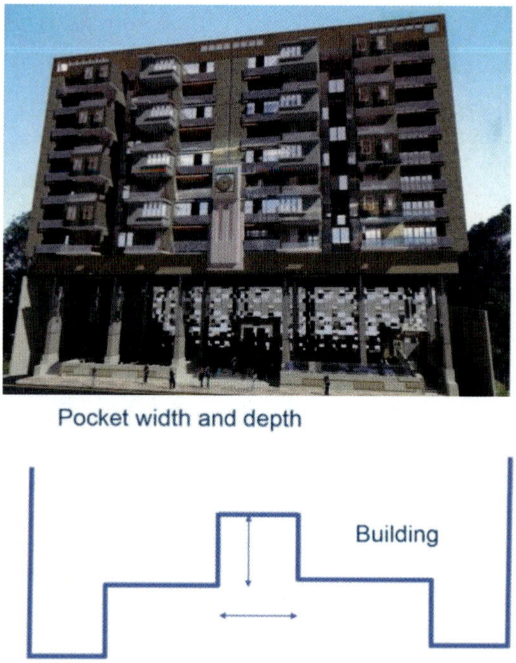

Fig. (6.8). Light pockets.

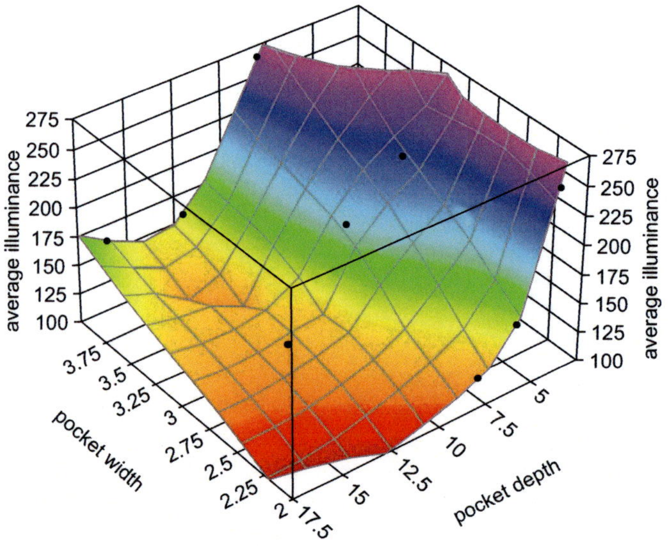

Fig. (6.9). Effect of pocket width and depth on the average illuminance inside the adjacent rooms.

Mirrors have also been used to redirect light into urban spaces. An example is shown in Fig. (**6.10**) where mirrors were placed to redirect light to a village square. The village square had never seen direct sunlight due to its location between two mountains. Another example is shown in Fig. (**6.11**) where a building in Spain was equipped with mirrors to redirect light downwards to the deep dark space between two adjacent buildings.

Fig. (6.10). Mirrors used to reflect light inside a village square. http://edition.cnn.com/2013/11/01/travel/rjukan-sun-mirror/.

A new sine-wave panel that redirects light downward was developed. The panel was made from a synthetic transparent thermoplastic material, Polymethyl Methacrylate (PMMA), which is readily available under commercial names such as Plexiglas or Acrylic.

Fig. (6.11). Mirrors used to reflect light downward to a building. https://en.wikipedia.org/wiki/Green_building.

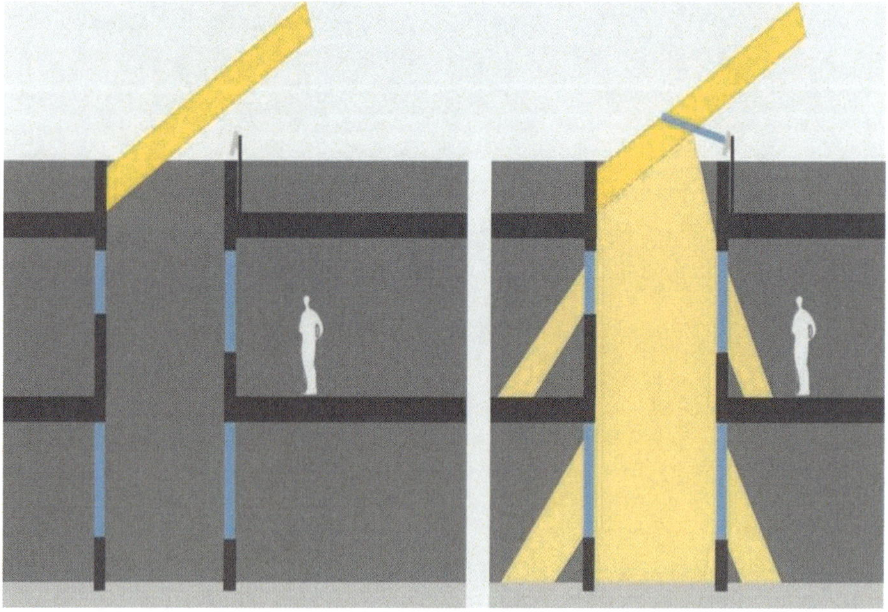

Fig. (6.12). The concept of light redirection.

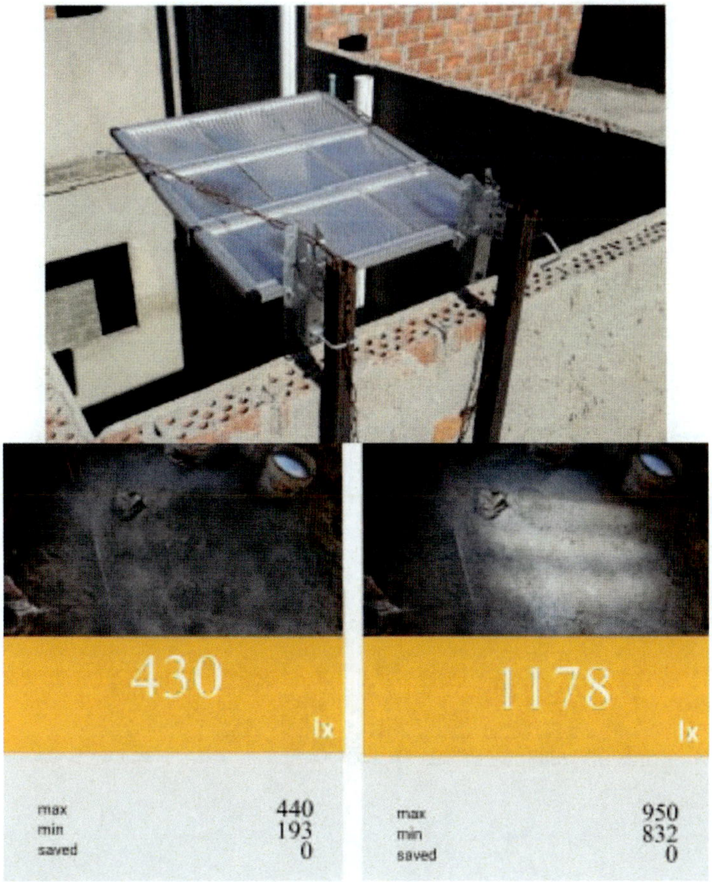

Fig. (6.13). A light redirecting panel.

Solar tubes is another technology that is capable of transporting light to spaces without direct access to the sky or the solar beam. Light tubes, which are sometimes referred to as a light pipe (also known as a sun scope, sun pipe, tubular daylighting device, or daylight pipe), are light redirection devices used for transporting and distributing daylight.

Light tubes have recently acquired significant market diffusion with implementation in many commercial and residential projects worldwide. Consequently, there are many commercial companies that provide proprietary light tube systems with different designs. There are essentially two types of solar tubes: the hollow type which transports light by reflection and the transparent solid type which transport the light by total internal reflection. Almost all daylighting application utilize the first type. Some are even equipped with LED lighting and a light sensor to provide a uniform amount of lumens throughout the day.

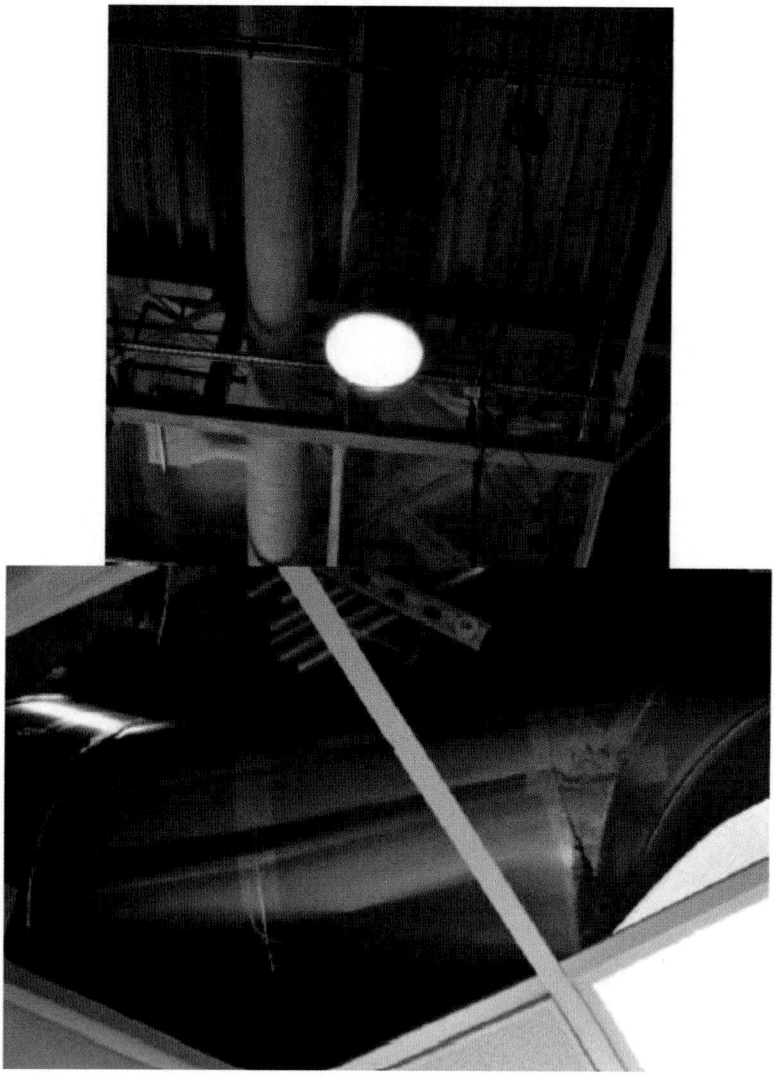

Fig. (6.14). Solar tubes.

QUESTIONS

1. Research the relationship between daylight and rickets.

2. Which of the different sky models discussed is more versatile?
3. Research the equation for the CIE sky model and explain the different terms involved.
4. What is the difference between point-by-point calculations for daylighting *versus* ray tracing?
5. What are the advantages of mirror light redirection based systems over light refraction based systems such as the panels in Fig. (**13**)?
6. What are the dis-advantages of mirror light redirection based systems over light refraction based systems such as the panels in Fig. (**13**)?

REFERENCES

[1] 2010 U.S. Lighting Market Characterization, "Solid-State Lighting Program, Building Technologies Program, Office of Energy Efficiency and Renewable Energy", U.S. Department of Energy.

[2] M. Boubekri, *Daylighting, Architecture and Health: Building Design Strategies.*. 1st ed Elsevier Ltd., 2008.

[3] G. Loisos, "Daylighting in Schools: An Investigation into the Relationship between Daylighting and Human Performance", 1999.

[4] P. Plympton, S. Conway, and K. Epstein, "Daylighting in Schools: Improving Student Performance and Health at a Price Schools Can Afford", *American Solar Energy Society Conference,* 2000

[5] L. Heschong, D. Mahone, K. Kuttaiah, N. Stone, C. Chappell, J. McHugh, J. Burton, S. Okura, R. Wright, and B. Erwin, *Skylighting and Retail Sales: An Investigation into the Relationship between Daylighting and Human Performance.* 1999.

[6] S. Aripin, "Healing Architecture': Daylight in Hospital Design", *Conference on Sustainable Building South East Asia,* vol. 5, p. 7, 2007.

[7] I. E. S. of North America; Daylight Metrics Committee, *Approved Method: IES Spatial Daylight Autonomy (sDA) and Annual Sunlight Exposure (ASE).*. Illuminating Engineering Society of North America, 2012.

[8] R.L. Knowles, *Energy and Form: An Ecological Approach to Urban Growth.*. MIT Press: Cambridge, MA, 1974.

[9] R.L. Knowles, *Sun Rhythm Form.* MIT Press: Cambridge, MA, 1982.

[10] K.E. Kensek, and R. Knowles, "Work in Progress: Solar Zoning and Solar Envelopes", *ACADIA Q.,* vol. 14, no. 2, pp. 11-17, 1995.

[11] R. Knowles, and V. Marguerite, "Solar Access and Urban Form", *AIA J.,* vol. 69, no. 2, pp. 42-49, 1980.

[12] Ž. Kristl, and A. Krainer, "Light Wells in Residential Building as a Complementary Daylight Source", *Sol. Energy,* vol. 65, no. 3, pp. 197-206, 1999.

[13] N. Lechner, *Heating, Cooling, Lighting: Sustainable Design Methods for Architects.* 2nd ed John Wiley

& Sons: New York, NY:, 2001.

[14] Y. Su, H. Han, S.B. Riffat, and N. Patel, "Evaluation of a Lightwell Design for Multi-Storey Buildings", *Int. J. Energy Res.,* vol. 34, no. 5, pp. 387-392, 2010. [http://dx.doi.org/10.1002/er.1651]

[15] B.M. Aly Ahmed, K.M. Nassar, and A. Asr, "Parametric Study of Light-Well Design for Day-Lighting Analysis under Clear Skies", *IACSIT Int. J. Eng. Technol.,* vol. 6, no. 1, pp. 81-85, 2014. [http://dx.doi.org/10.7763/IJET.2014.V6.670]

CHAPTER 7

Challenges in The Arab World

Sarah Mousa[2,#], **Mohamed El-Morsi**[1] and **Salah El-Haggar**[1,*]

[1] *The American University in Cairo, Department of Mechanical Engineering, Cairo, Egypt*

[2] *Arab Studies and Development, 5 Fox Glove Court, Wynantskill 12198, New York, USA*

Abstract: The MENA region receives between 22% and 26% of all solar energy striking the earth. This translates to a potential for solar energy per square kilometer per year equivalent to the energy generated from 1 to 2 million barrels of oil. Despite this, the region continues to deplete limited resources and escalate tensions over competition for the use of these resources. The MENA region only holds 1.8% of the world installed capacity of solar water heaters and 0.68% of the installed PV worldwide capacity. Governments of the Arab world often don't regard environmental protection and associated methods, renewable energy, as an economically or developmentally important. As a result the effort is often unfocused and driven simply by modest concern for international treaties. This chapter presents some of the challenges facing the implementation of solar powered technologies in the Arab world. These challenges are related to both the technologies themselves and the political and social factors.

Keywords: Arab world, Battery, Efficiency, Obstacles, Passive solar buildings, Photovoltaic, Solar cookers, Solar dryers, Solar electricity, Storage, Water heaters.

INTRODUCTION

The Arab world possesses great potential for renewable energy; the vast amount of available solar energy and land that can be exploited makes the region an ideal location for the implementation of solar energy technologies. Despite this, the

* **Corresponding author Salah El-Haggar:** Department of Mechanical Engineering, School of Sciences and Engineering, The American University in Cairo, AUC Avenue P.O. Box 74, New Cairo 11835, Egypt; Email: elhaggar@aucegypt.edu.

Founder of Shamsina

Salah El-Haggar, Sarah Mousa & Mohamed El-Morsi (Eds.)
All rights reserved-© 2016 Bentham Science Publishers

region continues to deplete limited resources and escalate tensions over competition for the use of these resources. As the region continues to consume fossil fuels, it is foremost harming its own land and populations as the resultant climate change and associated rise in sea level will first and foremost affect countries on the Mediterranean.

The implementation of solar energy technologies discussed in this book is a viable alternative to current energy sources. Solar energy presents itself as an inexhaustible resource which does not harm the environment. Once technologies are in place, a sustainable source is provided. Nonetheless, there are several challenges on geographical, technological, economical, institutional, and socio-cultural levels which are obstacles to solar energy usage and must be addressed.

AVAILABILITY OF SOLAR ENERGY

Solar energy is not necessarily a suitable energy source worldwide. As discussed in Chapter 1 and illustrated in Fig. (**1.1**), solar energy availability varies greatly depending on geographic location. The availability of solar energy also varies in a particular location depending on the season of the year and time of day. Atmospheric effects also discussed in Chapter 1, such as cloud coverage, could also limit the availability of solar energy. These factors mean that not all locations are ideal for the harnessing of solar energy. The Saharan Desert, both for its proximity to the equator and clear desert skies, is one of the most optimal locations for solar energy. Other areas with close proximity to the equator are also appropriate for the implementation of solar energy technologies. Before considering whether or not solar power technologies are appropriate for a particular community, the local availability of solar energy must be taken into account. The Middle East is one of the most optimal locations for the harnessing of solar energy worldwide. The region's solar energy endowments position it to lead the world in the field of renewable energy.

ENERGY EFFICIENCY

The amount of available solar energy does not alone convey the usefulness of a solar powered technology in a particular location. Current technologies are not able to take full advantage of all available solar energy. The amount of solar energy that can be harnessed for usable heat or electricity by these technologies is often very limited and must be taken into account when deciding whether or not the device will be useful.

In terms of photovoltaic panels, for example, energy harnessed cannot be directly used. Solar panels generate direct current (DC), while most equipment requires alternating current (AC). An inverter is needed, and is typically around 90% efficient. From the amount of available energy the overall efficiency of photovoltaic panels is at highest 25% [1]. For other solar powered technologies, efficiency must be considered along with price. In terms of water heaters, for example, flat panel heaters are less expensive but are also less efficient than evacuated tube heaters. Passive heaters are less expensive and rely solely on solar energy, but are also less efficient than water heaters which incorporate pumps. Energy efficiency is one of the main factors which must be considered when selecting a solar powered technology for implementation.

STORAGE EFFICIENCY

Energy is not evenly available throughout a day, and is not always needed for immediate usage. Technologies typically generate the greatest amount of power at midday, when the sun is most directly available, but not necessarily when needed. Since solar energy garnered is variable at any given time, it can rarely be relied on to power equipment directly. Energy garnered by photovoltaic panels requires a battery that stores the energy and provides it at a constant rate to equipment when needed. Batteries often do not store all energy garnered by a system, and thus add to its inefficiency. Storage efficiency must be accounted for when considering the overall efficiency of a system [1]. Furthermore, the cost of a battery and the cost of its replacement when needed must be accounting for when considering the total cost of a system. Storage batteries become less efficient with usage, and have limited lifetimes [1].

COST EFFICIENCY

The cost of implementing solar powered technologies, *versus* more common fossil fuel powered technologies, must be considered. Often times, however, those who consider solar power overlook running costs of equipment. Both initial costs and running costs (including energy costs and maintenance costs) of systems must be considered and compared. Solar technology is often relatively expensive to implement, but once running is almost free and will, after a period of time, be more cost efficient that technologies that rely on fossil fuels [1].

The available solar energy at a specific location throughout a given day and year, as well as how much of this available energy can be converted to usable energy and stored, is important to consider in terms of cost efficiency. Initial and running costs alone are not sufficient to select a specific system, it must be guaranteed that the system will garner sufficient energy for a household or community's needs.

INSTITUTIONAL OBSTACLES

Governments of the Arab world often regard environmental protection and conservation as constraints to rather than integral parts of development. Sustainable development is often ignored in favor of short-term economic growth. Because environmental protection and associated methods, renewable energy, are not viewed as an economically or developmentally important, the effort is often unfocused and driven simply by modest concern for international treaties. The region follows formalities, signing treaties such as Kyoto and issuing its own through the Arab League. In reality, however, sustainability is given low priority and little funding. The biggest challenge in establishing institutions for sustainable development is the difficulty in integrating economic, social and environmental concerns [2].

Institutions do not place any particular body as clearly responsible for sustainable development. Various interested parties, including the government, the private sector, and NGOs are often disconnected and do not coordinate efforts. NGO activity in the region is not connected to the decision-making process–often times, NGO projects get funding, are implemented, and then fall into disrepair [2].

The transition to a green economy presents itself as an opportunity for the Arab world, given its ideal climate for both solar and wind energy. Limited water sources add pressure for conservation and desalination. The Arab world could emerge as a world leader in the field. The political changes that we have recently seen sweeping the Arab world imply a major shift in the regional development path. Governments are now forced to put at the forefront social interests and concerns, chiefly among them youth employment, the lack of which was a leading cause of the revolutions [2].

SOCIO-CULTURAL FACTORS

The regional cultures and sub-cultures and the wide range of social and geographical groups found in the Arab world must be accounted for when considering the implementation of solar powered technologies. The wealthy, poor, urban, rural and remote/desert communities of the region are each associated with varying appropriate technologies due to their different geographic, social, cultural and economic conditions.

Passive solar buildings discussed in Chapter 5, are associated, in a modern sense, with costly and large scale projects. The concepts utilized in the building, however, are mostly inexpensive and if integrated in the original construction of a building could potentially provide moderate temperatures and save energy costs. The concepts can be applied to a wide range of the population, and especially since they are associated with traditional architectural concepts, should not face cultural barriers for implementation.

Solar dryers are mostly applicable for a specific community: rural men and women involved in the gathering and drying of crops, fruits and herbs. The key to solar dryers is proving their value. The example in Chapter 5, which discussed the implementation of solar dryers in a Palestinian community, shows that once convinced of the efficiency of a system a community will be quick to use it. As long as a target community can be shown that solar dryers are inexpensive and provide an enhanced product in less time than traditional drying, it should be quickly and widely implementable.

Solar cookers face more social obstacles due to the relatively longer cooking time. As shown in Chapter 5, solar cookers are especially useful in situations of need. When a community has no or erratic access to fossil fuel power sources, solar cookers are not only favorable, but necessary alternatives. For wealthier social groups, it may be possible to offer portable solar cookers for camping or picnicking purposes.

The Arab world faces a pressing need for fresh water sources, a need that is arguably a main cause of instability in the region. The country also borders sea water sources and has abundant solar energy, making solar desolation an ideal method for deriving potable water. Many communities in the region may ironically be located on a coast and be associated with fresh water deficiencies. As long as costs are reasonable, which they are for small scale solar desalination models, it should not be difficult to convince coastal communities to utilize solar energy to meet their fresh water needs.

Solar powered water heaters are applicable to a wide range of social groups. The examples of Darb al-Ahmar and Manshayet Nasser, cited Chapter 5, showed the urban poor sections of Cairo to be readily willing to use the technology. The key of the success of this project is that the community was able to see that solar powered water heaters could indeed provide a basic need at a lower overall cost. The wide range of solar powered water heaters, with increasing efficiencies and costs, makes it a possibility for a wide range of social classes, with the wealthy opting for the more expensive but also more efficient models.

Solar electricity, on a small scale, can be relatively less expensive than other electricity sources and is a viable option on a household or small community level. Chapter 5 illustrates that solar electricity can be especially useful in the Arab world where energy sources are unavailable or erratic. Once again, it is basic need and proof that a solar technology can fulfill this need that will result in the use of solar equipment. As in the Zabaleen community, once families see the benefits of a technology from their neighbors they will also encouraged to use the same technology.

In order to encourage populations to use solar technologies, the technologies must first be readily available and second their benefits must be widely known. While the environmental friendliness of these technologies is a primary benefit, the examples discussed in Chapter 5 show that, especially among lower-income technologies, it is crucial to prove the financial and benefits involved in the use of these technologies. Members of the Darb al-Ahmar communities boasted about their solar powered water heaters and how they cause no harm to the environment, but many if not all of these same individuals would not have implemented the technology if they were not convinced of its efficiency and relatively lower cost. Media can play an important role in making the public aware of important technologies and their benefits, but most be aware of the different communities which it will address and their varying needs.

DISCUSSION

The implementation of solar powered technologies is associated with several challenges, relating to both the technologies themselves and political and social factors. The finiteness of fossil fuel energy sources as well as the harm they cause to both the environment and humanity makes current energy traditions unsustainable. This adds to the urgency of exploring solar energy, despite the challenges. Furthermore, the changing political climate and the constant developments in the field increase potential for implementation of solar technologies in the Arab world. This book has made clear the need for exploration of solar technologies and the ease with which many of these technologies can be applied. It is thus upon the institutions and societies of the Arab world and elsewhere to begin making sustainable choices that are friendly to both environment and humanity and implement and improve upon existing technologies.

QUESTIONS

1. What are the geographical and seasonal limitations of solar energy?
2. What are the most significant technological challenges associated with solar energy technologies?
3. What are the costs of implementing solar energy technologies?
4. What are the political challenges of implementing renewable energy technology? Why are governments hesitant to implement renewable energy technologies?
5. What are the social and cultural challenges for the implementation of solar energy technologies? What are some reasons that a community might reject a particular technology?
6. Why is it important to begin implementing solar energy technologies?

REFERENCES

[1] M. Boxwell, "Solar Electricity Handbook: A Simple, Practical Guide to Solar Energy - Designing and Installing Photovoltaic Solar Electric Systems",

[2] "Regional Review of Institutions for Sustainable Development in the Arab Region," Economic and Social Commission for Western Asia (ESCWA)", In: *Economic and Social Commission for Western Asia (ESCWA).* United Nations: New York, NY, 2011.

CHAPTER 8

Feasibility of Solar Energy

Ghada Abdel Aziem[1], **Khaled Nassar**[1] and **Salah El-Haggar**[2,*]

[1] *The American University in Cairo, Department of Construction Engineering, Cairo, Egypt*

[2] *The American University in Cairo, Department of Mechanical Engineering, Cairo, Egypt*

Abstract: A model is developed to study the economic feasibility of adopting two kinds of solar technologies namely; solar water heaters and photovoltaics to substitute traditional systems in the residential sector. The model is a decision support system tool that guides investors and decision makers on the optimal scheme to utilize their funds and maximize the return on investment using an optimization approach over a duration of 10 years. The model considers offering a subsidy to support the diffusion of the systems through an initial fund of $1,500,000 in terms of a line of credit. The model computes the net present worth of the project and the line of credit payback period. In addition, the model provides annual percentages of the amount of money to be used in the reinvestment versus the amount used to pay back the line of credit. A case study is conducted and three proposed scenarios are applied namely; lease, loan and power-cuts tariff. The results of the study shows that the SWH line of credit payback period for the Lease, Loan and Power-Cut Tariff are 10, 9, and 10 years while the PV line of credit payback period for the Lease, Loan and Power-Cut Tariff are 8, 5, and 5 years.

Keywords: Decision support system tool, Economic challenges, Energy bond, Feed-in tariffs, Grants, implementation strategies, Lease, Line of credit, Loan, Net present worth, Optimization, Payback period, Photovoltaic panels, Power cut tariff, Rebates, Solar energy, Solar water heaters, Subsidy, Sustainable systems adoption, Tax reduction.

* **Corresponding author Salah El-Haggar:** Department of Mechanical Engineering, School of Sciences and Engineering, The American University in Cairo, AUC Avenue P.O. Box 74, New Cairo 11835, Egypt; Email: elhaggar@aucegypt.edu.

Salah El-Haggar, Sarah Mousa & Mohamed El-Morsi (Eds.)
All rights reserved-© 2016 Bentham Science Publishers

INTRODUCTION

The main purpose of this chapter is to discuss the economic feasibility and encourage the implementation of sustainable technologies in the field of solar energy in order to achieve more sustainable residential communities hence reducing carbon emissions as well as meeting the required electricity demand. There are a number of political, sociological, technical and economic obstacles that hinder the adoption of solar technologies. This chapter presents an optimization model that enables decision makers and professionals to study the economic feasibility of diffusing solar water heaters and photovoltaic systems.

ECONOMIC BARRIERS OF SOLAR TECHNOLOGIES DIFFUSION WORLDWIDE

The economic barriers are considered to be a major obstacle to the deployment of solar technologies. Timilsina *et al.* [1] stated that one of the major economic barriers is the high initial cost of solar technologies in comparison with current conventional technologies. It is mentioned as well that financing solar technologies is a critical barrier as most of the financing institutions avoid supporting solar technologies as a result of its high economic risks and long payback periods. Other economic barriers include the increase in the costs of solar technologies basic materials such as copper used in solar water heating systems manufacturing. Chedid [2] argues that the main economic constraints that may prohibit achieving solar water heaters diffusion in Lebanon include lack of proper funding, lack of government support in terms of tax deductions and incentive programs and high initial cost of the systems. Timilsina *et al.* [3] stated that the main barrier to promoting solar technologies in Thailand is its high initial cost. Veeraboina and Ratnam [4] claimed that in India the challenges include the low prices of natural gas, insufficient financing mechanisms that support renewable technologies and high initial costs of green technologies. According to Mills and Schleich [5] high initial costs of the solar water heating systems, low incomes of households, and rental of houses instead of ownership are considered to be the main barriers in Greece. Faiers and Neame [6] indicated that solar technologies will not be able to compete with current conventional technologies unless the electricity prices increase or the solar technologies become cheaper and more efficient. Ndzibah [7] has highlighted that based on interviews conducted with citizens from various regions in Ghana, most Ghanaians consider solar energy the most suitable alternative that can replace conventional energy however they cannot afford it.

POLICIES AND IMPLEMENTATION STRATEGIES

There are various policies that can support solar technologies. Timilsina *et al.* [1] mentioned different types of policies in their article "Solar Energy: Markets, Economics, and Policies" as follows: The feed in tariff (FIT) is suitable for all types of solar technologies. The renewable energy producers receive the cost of the energy generated in addition to their profits. It is claimed that the FIT can reduce the cost of renewable technologies as the potential renewable energy technologies guarantee long-term investments. The second policy is the investment tax credits which is an amount that can be deducted from the taxes paid by the tax payer in return of using solar energy technologies. Another policy is the renewable energy portfolio where governments set targets for percentages of the renewable energy supplied. To meet the targets energy producers and suppliers are required to provide certain percentages of renewable energy from the total amount of energy supplied. Suppliers who are not able to produce renewable energy can buy from suppliers who can. The net metering is a system that allows residential and commercial buildings to sell the extra solar energy they produce to the grid and this is implemented in Italy, Spain, Canada, USA and Australia. Another form of policies and solar technologies implementation strategies are the subsidies. Subsidies can be in terms of grants, soft loans, and capacity payments. In addition to the previous mandatory laws can force new construction and major renovation projects to use solar technologies. Finally, financing facilitation can be provided to support deployment of solar technologies. Timilsina *et al.* [3] mentioned that overcoming the economic obstacles of solar technologies can be implemented through Demand Side Management programs (DSM). Veerobina and Ratnam [4] have mentioned that the barriers of solar technologies adoption can be mitigated by developing suitable financial mechanisms consequently increase the demand of the renewable energy systems which can lead to a reduction in its costs. In addition to that the renewable technologies can be encouraged in poor and rural areas through government subsidies and innovative finance instruments to make it affordable for them. According to Zhai [8], renewable energy support policies include; Renewable Portfolio Standard (RPS), Feed in Tariff FIT, and tax and grant subsidies. Adhikari *et al.* [9] suggested that renewable energy projects shall be given a high priority to be implemented under the clean development mechanisms CDM which is one of Koyota protocol mechanisms. The CDM mechanism provides incentives to reduce CO2 emissions and achieve sustainable development. Drury *et al.* [10] stated that solar technologies diffusion can be implemented through a third-party that owns and operates solar technologies and sells the electricity generated to building occupants. Koroneos and Nanaki [11] have indicated that subsidies provided to support the use of renewable energy sources in addition to the increase in energy costs have led to a growth in the solar thermal energy market in many countries namely; Germany, Spain, Italy and France.

There are several methods to categorize solar technologies policies. Solar technologies adoption policies can be categorized into cost setting and quantity forcing policies, market facilitation and public investments and cost reduction policies. They can be also categorized according to their geographic and administrative scales, [8].

Case Studies

Recently, some governments have initiated incentive programs and policies to encourage the diffusion of solar water heating systems in their countries. The two main types of policies are subsidies and mandatory installation of SWH in new buildings. Table **8.1** is an overview of some of SWH policies and subsidy programs in various countries. The Indian government provided soft loans through a number of financial institutions to encourage the deployment of SWH systems. The loans' interest rates ranges between 2 to 5% according to the different categories of users and it is 0% for domestic users in a certain states. Other policies in India include mandatory solar water heating systems installation in new buildings in addition to providing rebates in electricity bills to users of solar water heating systems, [4]. Cyprus suffers from a shortage in the natural resources required to satisfy its energy demands. In order to support energy conservation investments and the replacement of the conventional with renewable energy resources, the Cypriot government has provided grants that represent around 30% of the total costs. As a result, the rate of evacuated tube and flat-plate collectors installation in Cyprus is the highest in the world, [12]. Germany and Austria provide subsidies that reach half of the installation costs [13], China produces solar water heating systems that are cheaper than western countries by about 80%. In addition to that the high efficiency of the evacuated tubes which allow the system to function in adverse weather conditions has led to the popularity of Solar Water Heating systems in China and about 30 million households in China have installed one solar water heating system [14]. In Greece 75% of the SWH investment costs can be deducted from the households' taxable incomes [13]. Spain was the second country to make the installation of solar water heating systems mandatory for new buildings in 2006 [14].

Table 8.1. Solar technologies polices worldwide.

Country	Policy/Initiative	Amount	Adoption
India	Soft Loans	-	
India	Mandatory in new buildings	N/A	
India	Rebates in bills for SWH users	-	
Cyprus	Grants	30% of total costs	Highest rate of SWH[2] installations in the World
Germany	Subsidy	Up to 50% of costs	
Germany	Subsidy (1999-2010)	60 to 125 €/m^2	
Austria	Subsidy	Up to 50% of costs	
China	Produces SWH cheaper	80%	30 million households
Greece	SWH costs are deducted from taxable incomes	Up to 75%	
Spain	Mandatory	N/A	
Taiwan	Incentive program for glazed flat plate (1986-1989)	2000 NTD[3]/m^2	58000 SWH installations
Taiwan	Incentive program for un-glazed flat plate (1986-1989)	1000 NTD/m^2	
Taiwan	Incentive program for glazed flat plate (started in 2000)	1500 NTD/m^2	
Taiwan	Incentive program for evacuate tube (started in 2000)	1000 NTD/m^2	

(Table 8.1) contd.....

Country	Policy/Initiative	Amount	Adoption
Taiwan	Subsidy	Up to 50%	
Switzerland	Subsidy	Up to 50%	
France	Tax credits	50% of installation costs	
Argentina	Subsidy provided by World bank and GEF	$30 million loan & $1 million grant	70,000 households & 1100 institutions

The German government provided subsidies that range between 60 to 125 E/m^2 between 1999 and 2010 [12]. The government grant is up to 50% in Kaohsiung in Taiwan and Basel in Switzerland [12]. The French government provides tax credits of about half the SWH installation costs [13]. The government of Taiwan has initiated two incentive programs. The first program was from 1986 to 1989 with a subsidy of 2000 NTD/m^2 and 1000 NTD/m^2 for glazed flat plate collectors and unglazed flat plate collectors respectively then the subsidy was reduced by 50%. This program resulted in the installation of 58000 solar water heating systems. The second program started in 2000 with a subsidy of 1500 NTD/m^2 and 1000 NTD/m^2 for glazed flat plate and evacuated tube solar collectors and unglazed flat plate collectors respectively [13]. The Argentinian government have launched Renewable Energy Project for the Rural Electricity Market (PERMER) in 1999 [15]. The target of the project is to provide electricity for 70,000 rural households and 1100 institutions from local renewable energy sources.

The Indian government has set a number of policies and has initiated various programs in order to benefit from India's high potential solar radiation [16]. The National Electricity Policy 2005 states that the percentage of electricity generated from renewable sources shall be increased and to issue a tender to obtain competitive prices.

SOLAR TECHNOLOGIES ADOPTION FACTORS

A number of studies have been conducted to figure out the factors that affect the potential customer's attitude towards the adoption of residential solar technologies. Drury *et al.* [10] have identified that the factors that affect the adoption of PV systems include occupant's age, household income, level of education, investment priorities, risk associated with using solar technologies, and neighborhood effects. Having analyzed the trends of PV adoption in Los Angeles and orange counties (USA) based on historical PV adoption and population characteristics, the results show that to increase the adoption of PV systems, customer owned PV is more suitable for areas with low income level while third-party owner ship PV products are more suitable for areas with high income level.

Although the government in the UK have initiated a program that supports two types of solar technologies namely solar thermal and solar photovoltaic systems through providing grants that decrease the costs of the systems by 50% few households have purchased the systems consequently; Faiers and Neame [6] presented a research to identify the domestic solar technologies adoption factors and barriers in the United Kingdom. Table **8.2** shows the process that a potential consumer shall go through to decide whether to adopt solar technologies or not. The process starts by being knowledgeable about the technology, then being interested in raising their awareness about it and comparing it with other technologies, followed by taking the decision to adopt the technology, then the implementation of the technology and finally confirming that the technology is meeting the expectations of the adopter. Adopters are categorized according to the timing of taking the decision of adopting a technology into innovators, early adopters, early majority, late majority, and laggards.

Table 8.2. The three key components of diffusion theory [6].

The Innovation – Decision process					
1. Knowledge	2. Persuasion	3. Decision		4. Implementation	5. Confirmation
Adopter Categories					
1. 'early adopters' (2.5%)	2. Early Adopters (12.5%)	THE CHASM	3. Early Majority (35%)	4. Late Majority (35%)	5. Laggards (15%)
A ttributes of Innovations					
Relative advantage	Compatibility	Observability		Trialability	Complexity

A number of interviews and surveys have been conducted to investigate the early majority consumer's feedback towards solar technologies, the attitude differences between the early and later adopters' categories and identify the adoption parameters. The results show that the consumers have positive attitudes towards residential solar technologies, however for future potential adopters either the technologies shall be improved or the financing policies shall be reviewed. The second observation show that the early adopters will keep adopting solar technologies while early majority are unlikely to adopt it without a change in the solar power. Finally; the results show that the financial, operational and aesthetic characteristics of solar technologies can be enhanced through spreading the awareness among the potential consumers through marketing strategies.

Song and Hamori [17] have collected and analyzed data over 11 years from 1996 to 2006 to identify the factors that affect the diffusion of solar photovoltaic systems in Japan. The results show that the adoption factors include regional government policy subsidies, new housing investment, household income, and environmental awareness.

EGYPT INTO PERSPECTIVE

Current Power Usage in Egypt

While the vast majority of the Egyptian population does have access to electricity, the fossil fuel sources upon which the country relies are finite and cause tangible harm to both the environment and human health. The local environmental and health effects of fossil fuel usage are most strongly pronounced in the densely populated and underprivileged areas of Egypt. While there is some presence of solar power technologies in the country, energy derived from this source is overall insignificant.

Egypt primarily relies on oil, gas and hydroelectricity for energy. Egypt produces oil and natural gas, and is a member of the Organization of Arab Petroleum Countries. The country imports oil to fulfill its consumption needs, and has excess gas which it exports to other countries in the region. In 2012, Egypt produced 164 TWh of electricity, 76% of which is from gas, 15% from oil and 8% from hydropower and the rest from PV and wind [18]. Egypt has an almost entirely integrated electrical grid and 94% of the population enjoys access to electricity [19].

Discoveries of oil reserves in Egypt occurred throughout the 20th century—major oil fields are located in Sinai, the Suez and the Eastern and Western Deserts.

Egypt's oil reserves are estimated at around 3.7 billion barrels. Major discoveries of natural gas sources were made in Egypt in the 1990s. The country's reserves are estimated at 66 trillion cubic feet, while more deposits are expected to be discovered. Egypt exports natural gas to the Middle East through the Arab Gas Pipeline [20].

A third significant source of energy in Egypt is hydroelectricity (a form of renewable energy) garnered from the Aswan High Dam and Aswan Reservoir Dams across the Nile River. The dams were first used for the purpose of electricity in 1967, and presently accounts for approximately 8% of Egypt's energy needs. The Aswan High Dam was associated with projects to provide electricity to rural villages previously without it, and was largely successful in this respect. Fig. (**8.4**), illustrates Egypt's energy usage.

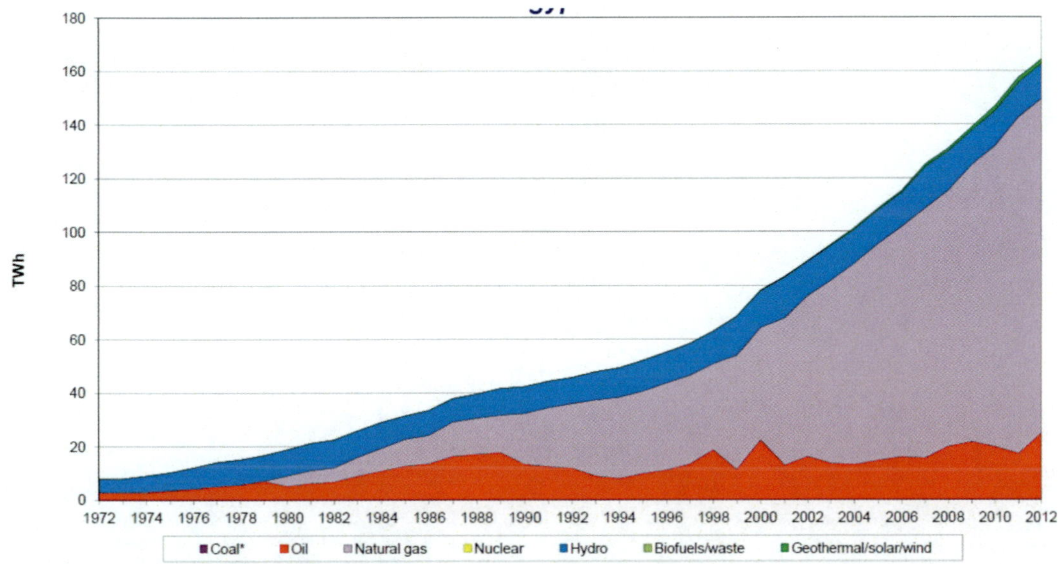

Fig. (8.1). Electricity generation by fuel [18].

Egypt's potential for harnessing energy from the sun is virtually untapped. There are currently several projects being planned to exploit the relatively high concentration of solar energy to which Egypt and the Middle East are exposed. Fig. (**8.2**), shows Egypt's solar energy potential [21]. DESERTEC [22] is a project that was recently proposed by several European countries to harness solar and wind energy from MENA on a large scale to be provided to Europe as well as MENA and other regions. While the project would overall be beneficial in its harnessing of clean energy sources, it is problematic for MENA on several levels. The project may be of some benefit to the region, but it will essentially be a project conceptualized and implemented by foreign rather than local entities. For Egypt and other MENA countries, it is dire to the sustainability of projects implemented in the region, as in any region, that they be locally led. Solutions to the numerous problems associated with fossil fuel usage in Egypt must come from the population itself for them to be truly effective and serve as a means to empower local development.

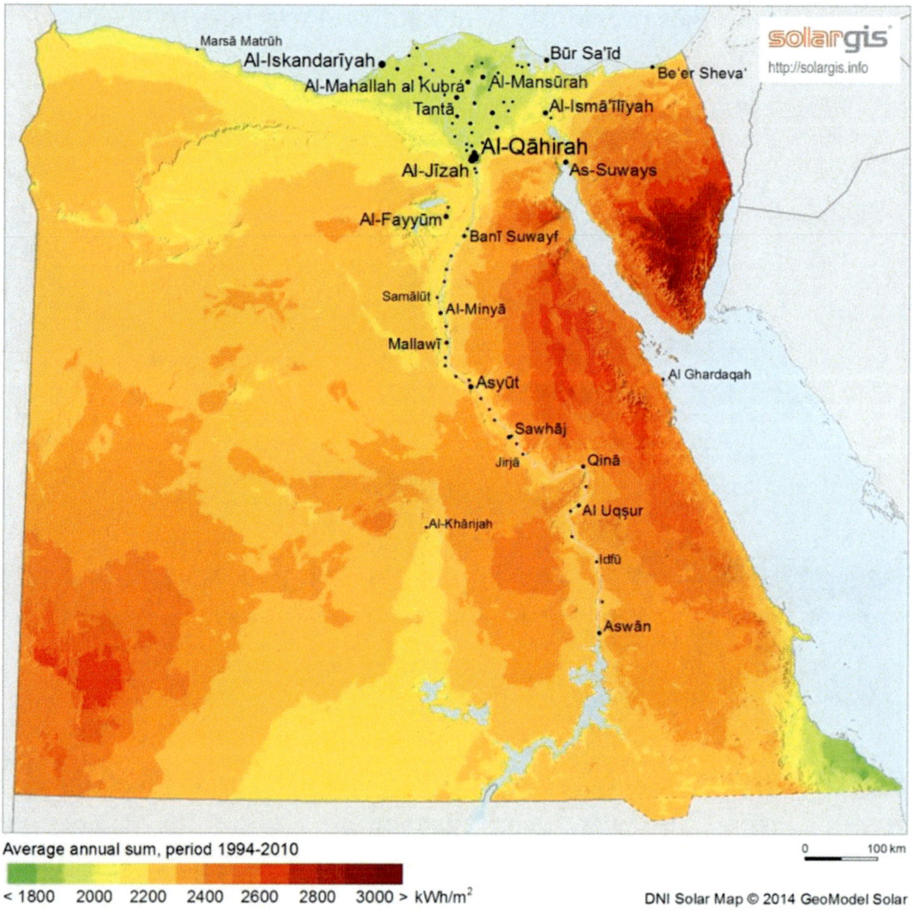

Fig. (8.2). Egypt's solar energy potential [21].

Egypt and Global Warming

All Arab countries, along with most of the world, have signed and ratified the Kyoto Protocol. Notable worldwide exceptions include the United States, which never signed, and Canada, which withdrew from the treaty. Egypt signed the Kyoto Protocol in 1999 and ratified it in 2005. In the context of Kyoto, Egypt took several steps to attempt to measure the release of greenhouse gases and to increase the use of renewable energy sources as an alternative to fossil fuels. Despite any efforts made, Egypt ranks 27 in a list of countries for top carbon dioxide emissions [20]. Fig. (**8.3**) shows a steady increase in Egypt's carbon dioxide emissions over time. Any temporal decreases seen in the CO_2 emissions below are likely correlated with a fall in economic activity.

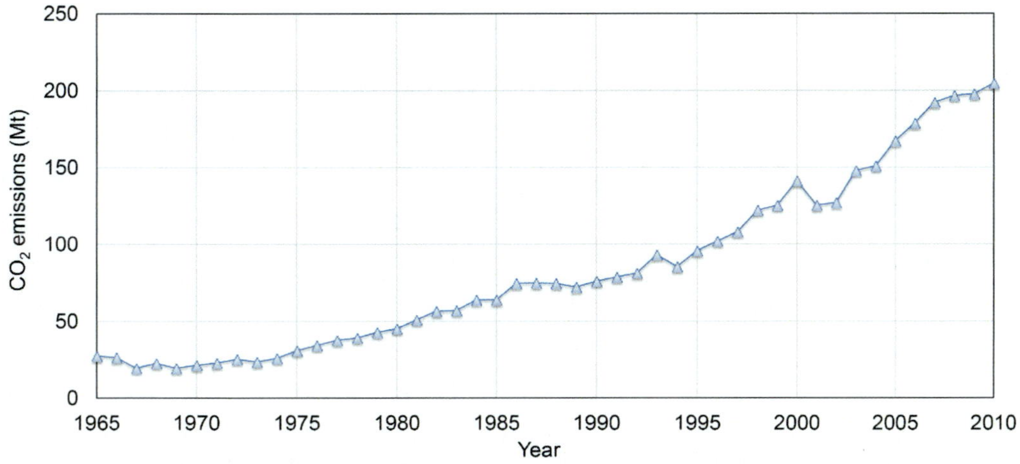

Fig. (8.3). Carbon dioxide levels in Egypt [23].

Over the past three decades, the Egyptian government has attempted to address global warming. In the 1980s, regulations were put forth requiring the use of solar powered technologies for all buildings constructed thereafter. The Egyptian government established the New and Renewable Energy Authority (NREA) in 1986 with the goal of introducing solar powered technologies on a commercial scale. In 2008 the NREA put forth a strategy to make renewable energy contributions 20% of energy in Egypt. The private sector was offered benefits in exchange for the use of solar powered technologies, including a void of custom duties on technologies [24]. In 2009, a solar-wind hybrid system was established in Kuraymat, 90 km outside of Cairo. The solar aspect of the plant relies on Concentrated Solar Power (CSP), and was functional starting late 2010.

Many of the plans put forth were not materialized and regulations were very poorly enforced, if any efforts at all were made to ensure their implementation. Fig. (**8.3**) confirms that CO_2 emissions have been steadily increasing in Egypt, and renewable energy contributions in the country are still relatively insignificant. While there have been several businesses established for the sale of solar-powered technologies in recent years, nearly all of these businesses import rather than locally manufacture technologies and awareness among potential consumers is null.[4]

Economic Barriers of Solar Technologies Diffusion

The diffusion of solar technologies in Egypt is facing a number of obstacles including political, sociological, economical and technical obstacles. Patlitzianas [25] summarizes the barriers In his article "Solar Energy in Egypt: Significant Business Opportunities" as follows. Direct subsidy to fuel prices is considered to be the main economic barrier to solar energy development in Egypt. It is claimed that in 2006, 8% of Egypt's budget was spent on the energy subsidies. Although electricity production requires 60% of the domestic natural gas and 15% of petroleum supply in Egypt, electricity prices in Egypt are one of the lowest in the world. One drawback of low electricity prices due to subsidies in Egypt is the overconsumption of energy. The average prices of the subsidized sales tariff do not cover the supply costs consequently the Egyptian Electricity Holding company is financially unstable. Also the Egyptian law forces gas producers to provide gas to the power plants at fixed rates that hardly cover the extraction and transportation costs. The Egyptian government has been trying to increase the electricity tariffs on a gradual basis to meet its actual generation costs. 2004 was the first time the prices of the electricity tariffs were increased since 1992 it was increased by 8.6% and it was planned that further increases will take place at a 5% rate. In addition to the low fuel cost in Egypt as result of subsidies there is a lack of support for solar energy investments in terms of financing solar projects. Solar technologies Investments that are of a small-scale nature rarely find a financial institution to fund it properly. Although the Egyptian government has aimed to achieve providing 20% of Egypt's electricity demand from renewable sources by 2010, all the decisions have been in favor of wind energy.

Another economic barrier is that because of the "Old Rental Law" in Egypt the owners of the units rented are not keen on reducing the electricity bills of their renters as the rents of the units cannot be increased by law. Finally, the high percentage of poverty in Egypt hinders the citizens from improving their social and economic conditions as they need to meet their basic needs first. The sociological barriers include that it takes time and effort to convince people with change and innovations. The consumers concerns about environmental impacts and climate change depend on their level of awareness and their level of education. Also some consumers think that PV systems can cause visual pollution on the top of their buildings.

Policies and Initiatives in Egypt

Parties Involved

According to Patlitzianas [25], the parties involved in the solar energy sector performance in Egypt are as follows:

1. MOEE: Ministry of Electricity and Energy
2. NERA: New and Renewable Energy Authority
3. NIS: National Institute of Standards
4. RCREEE: Regional Centre for Renewable Energies and Energy Efficiency
5. GOEIC: General Authority of Export and Import Control

In addition to the parties mentioned above there are other international parties involved in the Egyptian solar energy market

1. JCEE: Joint Committee on Renewable Energy and Energy Efficiency
2. JICA: Japan International Cooperation Agency
3. GIZ: Deutsche Gesellschaft für Technische Zusammenarbeit

In their annual report (2010-2011) the Egyptian New and Renewable Energy Authority (NERA) summarized their previous and future solar energy projects and initiatives as follows:

Previous Policies and Initiatives

1. In 1980 the ministry of electricity imported 1000 SWH systems with various capacities in addition to establishing the first local solar water heaters company. The number of SWH companies has increased to 12 companies with a total number of manufactured and assembled SWH systems of 350,000 systems in 2010.
2. Providing a 25% subsidy and operation and maintenance for four years to support spreading SWH in hotels and resorts located in Hurghada and South Sinai.
3. The implementation of a project for lighting using photovoltaic cells in two villages in Siwa Oasis. The project consists of lighting 100 households, 40 street lights, 1 school, 3 mosques and two clinics.
4. The execution of the first solar thermal power plant at Kuraiemat with total capacity of 140 MW financed by the global environment facility and Japan bank for international development. The operation of the solar power plant started in 30th of June 2011.

Future Initiatives

1. Two feasibility studies have been conducted for the implementation of two 20 MW photovoltaic power plants in Hurghada and Kom Imbou in conjunction with the Japanese and the French governments respectively.
2. The implementation of four photovoltaic plants with a total capacity of 20 MW has been incorporated in the solar energy program five year plan (2012-2017).

ECONOMIC MODEL

The economic model proposes three scenarios (Lease, Loan and Power Cut Tariff) that can be applied using the photovoltaic PV and the solar water heater SWH models as shown in Fig. (**8.4**). The economic model development will be divided into the solar water heater model and photovoltaic model for each scenario.

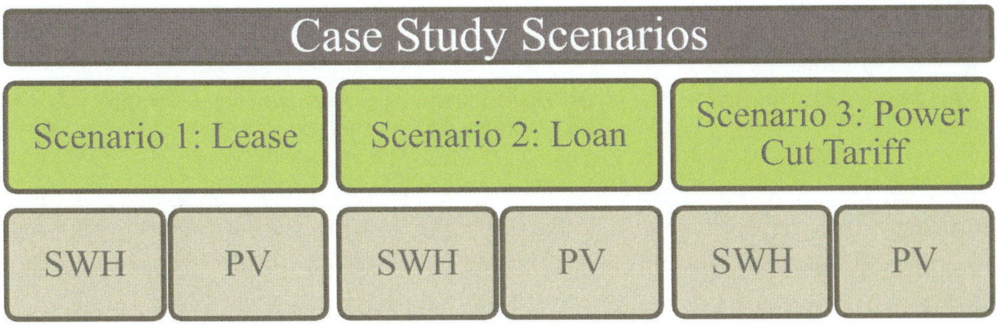

Fig. (8.4). Scenarios of the economic model.

Solar Water Heaters Model

The SWH model, shown in Fig. (**8.5**), is comprised of 5 modules namely; Project data, SWH System data, Electricity savings, Cash flow and Results.

Fig. (8.5). Solar water heater model.

Module 1: Project Data

The Project Data module, shown in Fig. (**8.6**), is designed to calculate the potential number of solar water heating systems installed based on a number of input parameters determined by the user such as: the Fund invested, the targeted number of households and the percentage of subsidy provided to facilitate the adoption of the systems. The systems' adoption percentages vary according to the subsidy percentage provided. The interest and inflation rates are used to estimate the increase in the costs of the systems as well as the electricity savings in the following modules.

Inputs and Outputs

The input and output parameters required for this module, shown in Fig. (**8.6**), are:

1. **Fund**: The user has to identify the amount of fund that will be invested in the project. That fund shall cover the subsidized costs related to SWH systems initial and running costs such as operation and maintenance, customs, and taxes.
2. **Interest Rate**: The line of credit interest rate shall be defined.
3. **Inflation**: The percentage of inflation is required to estimate the annual increase in systems' costs.
4. **Number of Households**: The user shall determine the number of potential households targeted to adopt SWH.
5. **Subsidy Percentage:** The percentage of money of the SWH initial cost provided by the investor to the consumer. The subsidy percentage ranges between 0% and 100% depending on the user's target and the amount of fund available.
6. **Adoption Percentage**: The SWH adoption percentage is the percentage of potential households that will agree to adopt the solar system. It depends on the percentage of subsidy provided to support the adoption of the SWH. The user can either determine the adoption percentage or use the default percentage provided by the model.
7. **Cumulative Number of Systems installed over 10 years**: The number of installations in the first year, $I_{yr\ 1}$, is calculated based on the total targeted population, P, and the adoption percentage, A, that depends on the subsidy percentage, as shown in Equation (8-1).

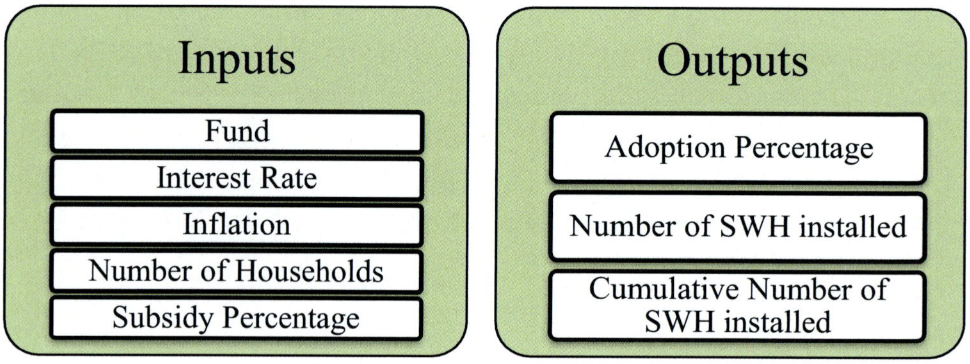

Fig. (8.6). MODULE 1: PROJECT DATA parameters.

$$I_{yr\,1} = A \times P \tag{8-1}$$

The first year installations are deducted from the total targeted population to calculate the number of installations left from the first year. This is performed for every year, as shown in Equations (8-2) and (8-3).

$$I_{rem\,yr\,1} = P - I_{yr\,1} \tag{8-2}$$

$$I_{rem\,yr\,2} = I_{rem\,yr\,1} - I_{yr\,2} \tag{8-3}$$

where $I_{rem\,yr\,1}$, is the number of installations left in the first year and $I_{yr\,2}$, is the number of installations in the second year. The cumulative number of installations, I_{cum}, is the summation of the number of systems installed every year, as shown in Equations (8-4) and (8-5). $I_{cum\,1}$ and $I_{cum\,2}$ are the cumulative number of installations in the first and second year, respectively.

$$I_{cum\,1} = I_{yr\,1} \tag{8-4}$$

$$I_{cum\,2} = I_{cum\,1} + I_{yr\,2} \tag{8-5}$$

Module 2: SWH System Data

The main purpose of the second module, shown in Fig. (**8.7**), is to estimate the total subsidized costs for installed collectors. The user can select the type of collector from the model database. The capacity of the system is estimated based on the average number of occupants per household. The total cost per collector include the initial cost per collector, the operation and maintenance costs, and taxes and customs if applicable. Having estimated the total costs per collector, the total subsidized cost for all systems installed is calculated using the subsidy percentage entered by the user in the first module. The model uses the inflation rate identified by the user to estimate the increase in system prices throughout the duration of the project.

<u>Inputs and Outputs</u>

The input and output parameters required for this module, shown in Fig. (**8.7**), are:

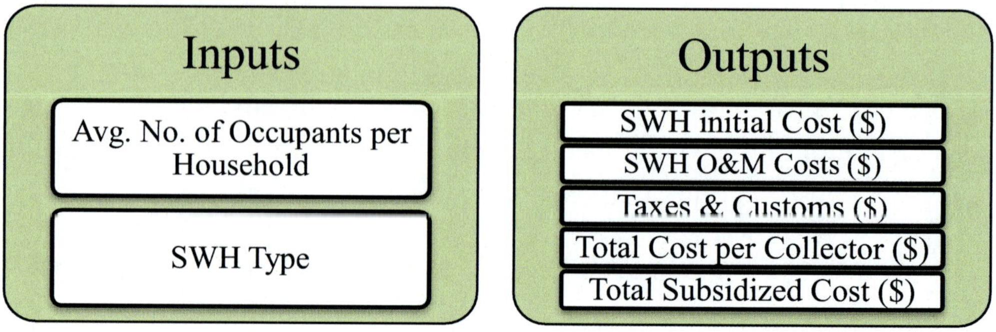

Fig. (8.7). MODULE 2: SWH SYSTEM DATA parameters.

1. **Average Number of Occupants per Household:** The number of occupants per household shall be defined by the user as it is used to calculate the average hot water daily consumption hence calculate the amount of electricity required to heat water. The average number of occupants per potential household is used to determine the capacity of the SWH.
2. **Solar Water Heater Type:** The user can select among the types of SWH systems incorporated in the SWH database. The types include local and imported flat plate and evacuated tube SWH.
3. **Total Cost per Collector:** The cost of the collector, *T c*, includes the initial cost, *I c*, operations and maintenance cost, *O M*, taxes, *T*, and customs, *C*, as shown in Equation (8-6). The total cost at any year *x* depends on the increase in systems' prices as a result of inflation, as shown in Equation (8-7).

$$Tc = Ic + OM + T + C \qquad \textbf{(8-6)}$$

$$Tc\ (x) = Tc \times (1 + inf\%) \qquad \textbf{(8-7)}$$

4. **Total Subsidized Cost per Collector:** The subsidized cost per collector, *S c*, is calculated using the subsidy percentage, *S*, as shown in Equation (8-8). The total subsidized cost, *T s c*, is calculated based on the subsidized cost per collector and the total number of installations in any year *x*, *I (x)*, as shown in Equation (8-9).

$$Sc = Tc \times S \qquad \textbf{(8-8)}$$

$$Tsc = Sc \times I\ (x) \qquad \textbf{(8-9)}$$

The cumulative subsidized cost, Csc, of collectors is calculated according to Equation (8-10).

$$Csc = Tsc\ (x1) + Tsc\ (x2) \ldots .etc \qquad \textbf{(8-10)}$$

5. **Solar Water Heating System Costs:** If applicable, taxes and customs are calculated as a percentage of the system's initial cost as shown in Equations (8-11) and (8-12).

$$T = 0.1 \times Ic \qquad \textbf{(8-11)}$$

$$\mathrm{C} = 0.05 \times Ic \qquad \textbf{(8-12)}$$

Module 3: Electricity Savings

The electricity savings module, shown in Fig. (**8.8**), calculates the total amount of electricity saved in terms of money as a result of replacing electric water heaters by solar water heaters. The model uses RETScreen [26] to calculate the amount of savings per collector based on a number of parameters including electricity tariff, daily hot water usage and SWH type and costs.

Inputs and Outputs

The input and output parameters required for this module, shown in Fig. (**8.8**), are:

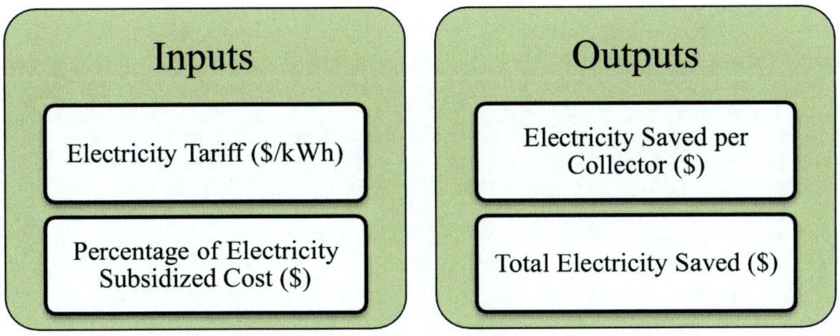

Fig. (8.8). MODULE 3: ELECTRICITY SAVINGS parameters.

1. **Electricity Tariff ($/kWh):** The electricity tariff is measured in $/kWh.
2. **Percentage of Electricity Subsidized Costs (%):** Replacing Electric water

heaters with solar water heaters is supposed to result in energy savings from a customers' point of view. From the government prospective the electricity savings will include both the electricity savings and the subsidy provided as a percentage of the electricity tariff.

3. **Total Electricity Saved ($):** The total amount of electricity saved per collector in any year *x*, *T es (x)*, as a result of using solar systems, is determined using the RETScreen software [26]. The software uses NASA meteorological database [27] in addition to a number of input parameters required to estimate the savings. Equations (8-13) and (8-14) depicts the calculation for the total amount of electricity saved in any year *x* and the cumulative savings, *T es(cum)*, respectively, where *Es(x)* is the electricity saved for any year *x*.

$$Tes(x) = Es(x) \times I(x) \quad \textbf{(8-13)}$$

$$Tes(cum) = Tes(x1) + Tes(x2) \ldots etc \quad \textbf{(8-14)}$$

Module 4: Cash Flow

The purpose of this module, shown in Fig. **(8.9)**, is to calculate the profits gained from the project on an annual basis in addition to the minimum payback period of the project and the percentage of money used to reinvest in the following year and the percentage paid for the loan.

Inputs and Outputs

The input and output parameters required for this module, shown in Fig. **(8.9)**, are:

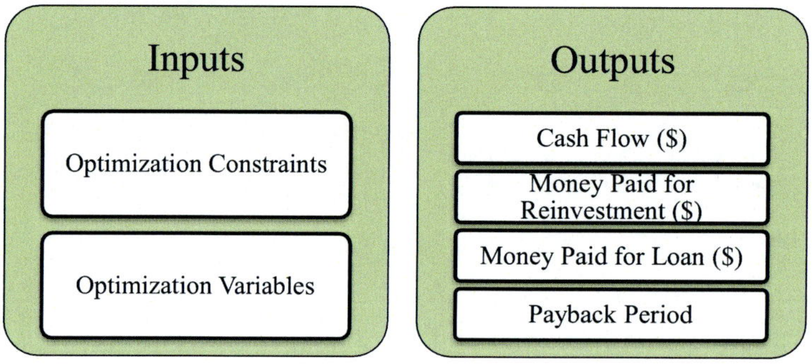

Fig. (8.9). MODULE 4: CASH FLOW parameters.

1. **Optimization Constraints:** The maximum profits are the main output of the model as previously mentioned. The profits gained are calculated by deducting the total amount of costs including SWH systems initial, maintenance and operation costs from the total amount of electricity savings. The profits are divided into the amount of money that will be used to fund the project the following year and the amount of money that will be paid as payback of the loan. The percentages of the two divisions are considered the model variables. The constraints include that the variables have to be positive values and have to be less than 1 and that the cumulative loan payment should not exceed the amount of the loan after adding the interest rate.
2. **Optimization:** The optimization is performed using solver and it is used to calculate the percentage of reinvestment and line of credit payback payment.
3. **Constraints**: The percentage of the cash flow savings in year x is added to the electricity savings as shown in Equation (8-15), where $T\,eso(x)$ is the optimized $T\,es(x)$.

$$Teso(x) = Es(x) \times I(x1) + Mf(x) \qquad \textbf{(8-15)}$$

4. **Calculation of the Cash Flow:** The cash flow in the first year, $CF(x1)$, depends on the available funds, F, total electricity savings during the first year, $T\,es(x1)$, and the total subsidized cost, $T\,sc(x1)$, as depicted by Equation (8-16). The cumulative cash flow is the summation of cash flows every year, as shown in Equation (8-17).

$$CF(x1) = F + Tes(x1) - Tsc(x1) \qquad \textbf{(8-16)}$$

$$CF(cum) = CF(x1) + CF(x2) \ldots etc \qquad \textbf{(8-17)}$$

If the cash flow result is positive the amount of money saved is divided using optimization into an amount to pay the loan, $M\,l$, and another to reinvest in the project, M f, according to Equations (8-18) and (8-19), respectively. $L\%$ and $F\%$ are the percentages of loan payment and money paid to reinvest in the project, respectively.

$$Ml = CF \times L\% \qquad \textbf{(8-18)}$$

$$Mf = CF \times F\% \qquad \textbf{(8-19)}$$

Photovoltaic Model

The PV model, shown in Fig. (**8.10**), is divided into 6 modules namely; Project data, appliances and energy demand, PV system data, Electricity savings, cash flow and results. The first five modules will be further explained in the following sections.

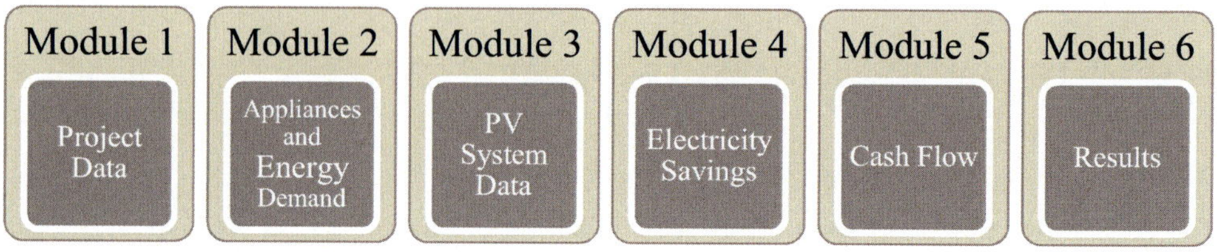

Fig. (8.10). Photovoltaic model.

Module 1: Project Data

The Project Data module, shown in Fig. (**8.11**), is designed to calculate the potential number of Photovoltaic systems installed based on a number of input parameters determined by the user such as the Fund invested, the targeted number of households and the percentage of subsidy provided to facilitate the adoption of the systems. The systems' adoption percentages vary according to the subsidy percentage provided. The interest and inflation rates are to be used in order to estimate the increase in the costs of the systems as well as the electricity savings in the following modules. The average area of the roofs shall be determined by the user to estimate the number of PV modules installed.

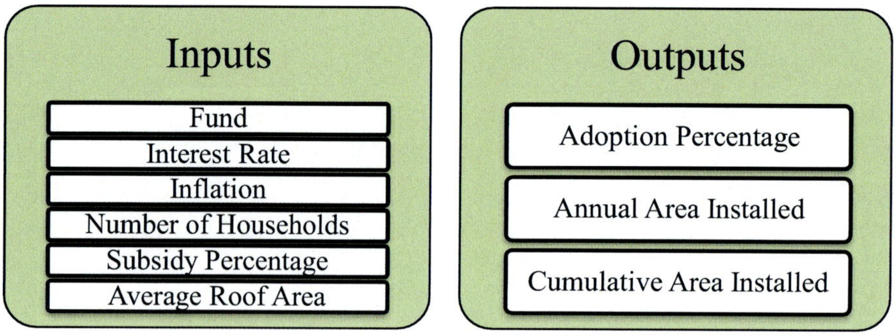

Fig. (8.11). MODULE 1: PROJECT DATA parameters.

Inputs and Outputs

The input and output parameters required for this module, shown in Fig. (**8.11**), include Fund, Interest Rate, Inflation, Subsidy and Adoption Percentages, and Cumulative Number of Systems installed over 10 years, are as described in section 0.

1. **Number of Households**: The user shall determine the number of potential households targeted to adopt PV systems. The targeted population shall be the citizens of a small city to a whole governorate depending on the amount of fund the user is intending to invest.
2. **Average Roof Area:** The average area of the roofs is required in order to determine the number of modules that can be installed.
3. **Percentage of Useful Area:** The useful area is the area where the PV modules can be installed and it is calculated by excluding the area of any obstacles or shaded areas from the total roof area.
4. **Annual Area Installed:** The annual area installed is the average useful area of the roofs of all the targeted households.
5. **The Cumulative Area installed:** The area installed in the first year, *A(ins)*1, depends on the roof area, *Ar*, percentage of useful roof area, *Au*, and the number of SWH installations in the first year, as presented in Equation (8-20). The cumulative area installed, $I_{cum\ 1}$, is calculated according to Equation (8-21).

$$A(ins)1 = Ar * Au * Iyr\ 1 \quad \textbf{(8-20)}$$

$$A(cum)1 = I_{cum\ 1} * Ar * Au \quad \textbf{(8-21)}$$

Module 2: Appliances and Energy Demand

The main purpose of this module, shown in Fig. (**8.12**), is to estimate the energy demand per household. The user is to identify the domestic appliances that shall use energy generated from PV modules instead of grid electricity. The energy required for the operation of the selected appliances is calculated based on the typical wattage identified by General Electric for each appliance. The user has to identify the number of hours and the number of appliances of each type. The user can select up to 15 Appliances including television, refrigerator, air conditioner and microwave.

Inputs and Outputs

The input and output parameters required for this module, shown in Fig. (**8.12**), are:

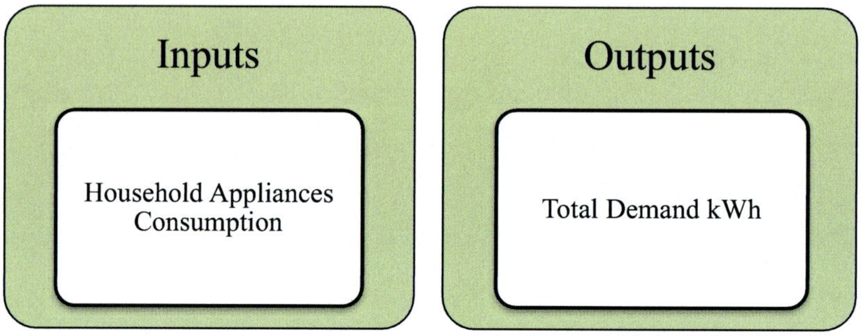

Fig. (8.12). MODULE 2: APPLIANCES AND ENERGY DEMAND parameters

1. **Household Appliances Consumption:** The user can select up to 15 different types of appliances to be operated using PV modules instead of conventional systems.
2. **Household Energy Demand:** The total demand of the selected appliances is calculated to estimate the electricity savings as illustrated in the fourth module. The total daily demand per household, *Tdh*, is estimated using Equation (8-22) based on typical appliance wattage, *W*, number of appliances, *N*, and number of hours of operation, *H*. The total annual demand, *Tad*, is estimated according to Equation (8-23), where *I(x)* is the number of installations in any year *x*.

$$Tdh = W \times N \times H \quad \textbf{(8-22)}$$

$$Tad = Tdh \times I(x) \times 365 \quad \textbf{(8-23)}$$

Module 3: PV System Data

The main purpose of this module, shown in Fig. (**8.13**), is to estimate the total subsidized cost of installed PV modules. The user selects the type of PV modules from a database that contains the initial and running costs of the system, the energy output of the module and the size of the module. The number of modules installed is calculated by dividing the useful roof area by the module size. The subsidized cost of modules installed and the energy produced are calculated to develop the cash flow in the following Modules. The model carries out a check to guarantee that the amount of energy generated from the modules is sufficient for the annual demand of the households.

Inputs and Outputs

The input and output parameters required for this module, shown in Fig. (**8.13**), are:

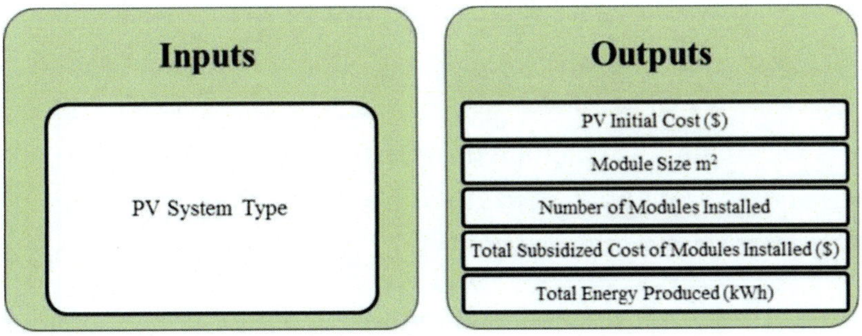

Fig. (8.13). MODULE 3: PV SYSTEMS DATA parameters.

1. **PV System Type:** The user selects the type of the PV system among a number of types incorporated into the model database. The database contains data about the initial and running costs and the size of the module.
2. **Module Size (m^2):** The module size varies from one type to the other and is used to estimate the number of modules that can be installed.
3. **Number of Modules Installed:** The number of modules installed, $I(x)$, is calculated by dividing the installed area, A_{ins}, which is the total households useful roof area by the area of the selected type of module, A_{PV}, as shown by Equation (8-24). The number of modules is used to estimate the total costs of the PV modules and the amount of energy generated.

$$I(x) = A_{ins} / A_{PV} \tag{8-24}$$

4. **Total Subsidized Cost of Modules Installed**: The total subsidized cost of modules depends on the percentage of subsidy provided.

5. **Total Energy Produced (kWh):** The total energy produced by the PV module, TE_{PV}, is estimated based on the type of PV system chosen and the amount of energy it generates, E_{PV}, as shown by Equation (8-25).

$$TE \quad = E \quad \times I(x) \times 365 \qquad \textbf{(8-25)}$$

The model carries out a check to verify that the energy produced by the modules can satisfy the household demand. If the cumulative energy generated, CE_{PV}, is less than the demand, the model advices to modify the selection. Else, the cumulative energy produced is used to estimate the energy savings and develop the model cash flow. The cumulative energy generated is calculated by summing the total energy produced by each PV module, as presented by Equation (8-26).

$$CE_{PV} = TE_{PV1} + TE_{PV2} + TE_{PV3} \ldots etc. \qquad \textbf{(8-26)}$$

The total cost per module is calculated according to Equations (8-6). The number of modules installed is estimated by dividing the total area installed per year by the area of the selected module, as shown by Equation.

The total cost of PV systems installed is calculated as shown by Equation (8-27).

$$T = TC \times I_{yr} \qquad \textbf{(8-27)}$$

The total subsidized cost of PV systems installed is estimated as presented by Equation (8-28).

$$TSC = T \times S \qquad \textbf{(8-28)}$$

Module 4: Electricity Savings

In the Electricity Savings module, shown in Fig. (**8.14**), the user identifies the electricity tariff and the percentage of subsidy provided for the electricity tariff to compute the annual electricity savings with and without the subsidy.

Inputs and Outputs

The input and output parameters required for this module, shown in Fig. (**8.14**), are:

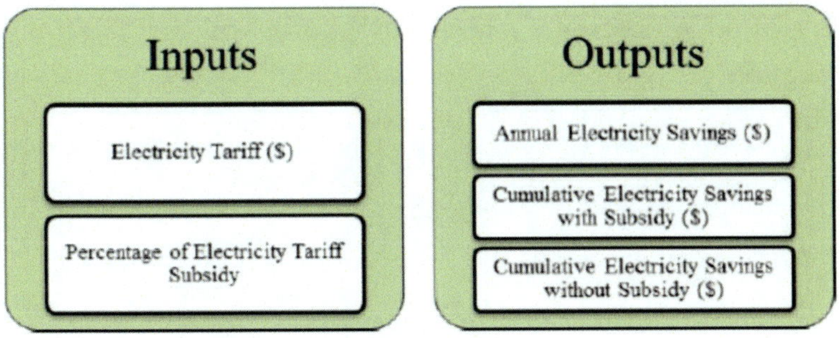

Fig. (8.14). MODULE 4: ELECTRICITY SAVINGS parameters.

1. **Electricity Tariff:** The electricity tariff is measured in $/kWh and is used to estimate the energy savings resulting from replacing conventional systems by PV systems.
2. **Percentage of Electricity Tariff Subsidy:** The percentage of electricity tariff subsidy is used to estimate the actual cost of electricity saved rather than the government subsidy.
3. **Annual Electricity Savings ($):** The PV system annual savings is estimated by the RETScreen software [26] and it is used to estimate the savings for all the systems installed.
4. **Cumulative Electricity Savings with Subsidy ($):** The subsidized cost of electricity tariff is used in this case to calculate the cumulative cost of electricity saved if the government subsidy is provided.
5. **Cumulative Electricity Savings without Subsidy ($):**The percentage of subsidy provided by the government for the electricity tariff is removed in this case to calculate the actual cost of electricity saved.

Module 5: Cash Flow

The purpose of this module, shown in Fig. **(8.15)**, is to calculate the profits gained from the project on annual basis in addition to the minimum payback period of the project and the percentage of money used to reinvest in the following year and the percentage paid for the loan.

Inputs and Outputs

The input and output parameters required for this module, shown in Fig. **(8.15)**, include Optimization Constraints, and Calculation of the Cash Flow, are as described in section 0.

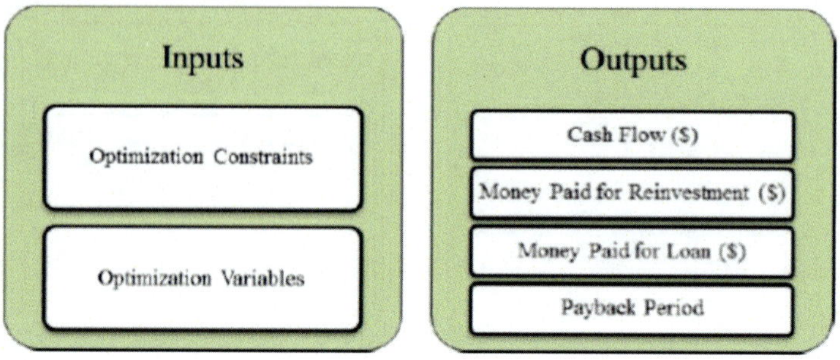

Fig. (8.15). MODULE 5: CASH FLOW parameters.

1. **Constraints:** The percentage of the cash flow costs in year *x* shall be subtracted from the PV systems costs, as shown in Equation (8-29).

$$TSC = T \times S + Mf(x) \tag{8-29}$$

CASE STUDY

The three proposed scenarios are applied to a gated residential compound in New Cairo – Egypt and all the electricity savings estimated by RETScreen [26] are based on Cairo climate data incorporated in the software. That location is chosen as the households' annual income is relatively high. Consequently, the residents can afford installing the solar systems in addition to the fact that their electricity consumption is relatively higher than other communities hence they are more willing to install solar systems. The targeted population for both SWH and PV systems is 2,000 households and the project is funded by a loan of $ 1,500,000 for each of the two types of systems with an interest rate of 10% over a period of 10 years. The basic scenario's subsidy percentage was determined to be 70% of the costs of the systems and it includes their initial costs, taxes and costumes as well as operation and maintenance costs while the adoption percentage is 50%.

Scenario 1: Lease

In this scenario, shown in Fig. (**8.16**), the investor funds the projects through a bank loan with an interest rate of 10%. The investor provides SWH & PV systems to the potential consumers with a subsidy of 70% of the systems' initial costs over a period of 10 years. The consumers shall install the systems on their rooftops and pay an annual fee for the lease of the systems through electricity bills with a fixed rate. The energy savings are estimated in order to determine the fixed amount paid to the investor. The model guides the investor on the amount of money paid to payback the loan and the amount used for the reinvestment of the project. The following sections show each module's input parameters and the results of the first scenario.

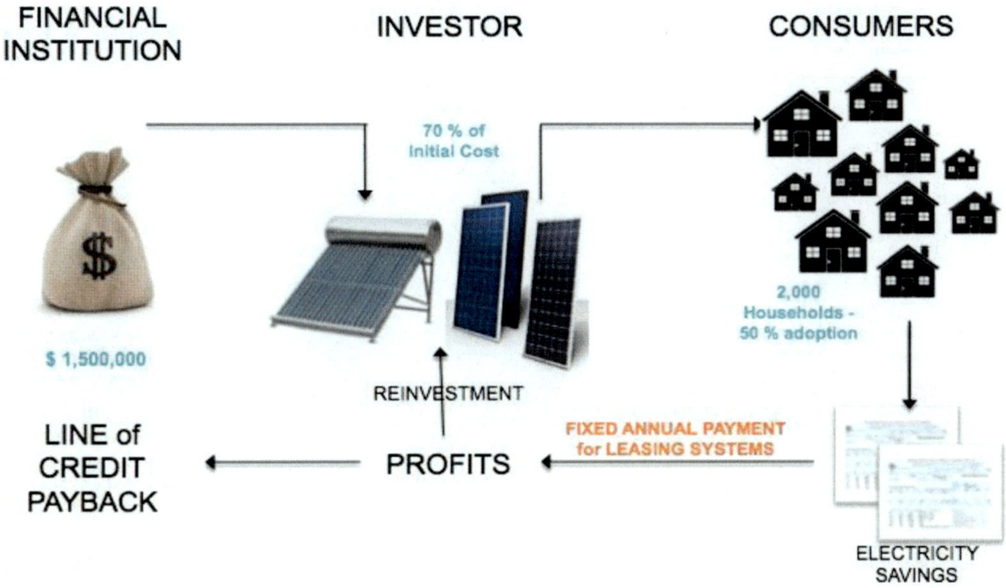

Fig. (8.16). Scenario 1: Lease.

SWH Model

<u>*Module 1: Project Data*</u>

<u>*Input Parameters*</u>

The input parameters for Module 1 are summarized in Table **8.3**. The number of households that adopted the SWH model in the first year is 1000 which is 50% of the targeted population and also it is the maximum number of systems that can be purchased by the fund provided, as shown in Table **8.3**.

In the following years the number of households that adopt the system decreases until it becomes 2 households in the 10th year. The reason behind this is that the model performs a check on the amount of money used to invest in the project in the previous year and verifies if it is sufficient to pay the subsidy for 50% of the targeted population in the current year or not. If that amount is not enough then the maximum number of subsidized systems that can be installed is shown instead. The cumulative number of systems installed over the total duration of the project is 1998 households.

Table 8.3. SWH – Lease: Module 1 - Project data.

Fund Amount in ($)	1500000									
Number of Households	2000									
Loan Interest Rate (%)	10									
Subsidy Percentage (%)	70									
Percentage Adoption (%)	50									
Parameters	**Year 1**	**Year 2**	**Year 3**	**Year 4**	**Year 5**	**Year 6**	**Year 7**	**Year 8**	**Year 9**	**Year 10**
Annual Number of Households Adopted	1000	500	250	125	62	31	16	8	4	2
Number of Households Adopted Left per Year	1000	500	250	125	63	31	16	8	4	2
Cumulative Number of Households Adopted	1000	1500	1750	1875	1937	1969	1984	1992	1996	1998

Module 2: SWH System Data

Inputs Parameters

The average number of occupants per household specified in the first scenario is 5 occupants, as shown in Table **8.4**. The type of SWH system selected is a local evacuated tube with a capacity of 50 liters and a total cost of $953. Only one system per household is installed, as shown in Table **8.4**. A 10% increase in the systems costs due to inflation is taken into consideration for the following years.

The total subsidized cost of the systems in the first year is $667,450. Table **8.4** shows that in some cases the total annual subsidized system costs is negative since the amount of money used to reinvest in the project is deducted from the costs of the systems in the preceding year.

Table 8.4. SWH – Lease: Module 2 - System data.

Parameters	Year 1	Year 2	Year 3	Year 4	Year 5	Year 6	Year 7	Year 8	Year 9	Year 10
Average Number of Occupants per Household	5	5	5	5	5	5	5	5	5	5
Average Daily Hot water Consumption (L/Day)	375	375	375	375	375	375	375	375	375	375
SWH Type	Evacuated Tube (Local) - 50 L									
Imported or Local	Local									
Number of SWH Systems per Household	1	1	1	1	1	1	1	1	1	1
System Initial Cost ($)	785	864	950	1045	1149	1264	1391	1530	1683	1851
SWH Operation & Maintenance Cost ($)	90	99	109	120	132	145	159	175	193	212
Taxes Percentage (%)	10	10	10	10	10	10	10	10	10	10
Customs Percentage	0	0	0	0	0	0	0	0	0	0.05
Taxes ($)	79	86	95	104	115	126	139	153	168	185
Customs ($)	0.00	0.00	0.00	0.00	0.00	0.00	0.00	0.00	0.00	92.55
Total Cost per SHW System ($)	954	1049	1154	1269	1396	1536	1689	1858	2044	2341

(Table 8.4) contd.....

Parameters	Year 1	Year 2	Year 3	Year 4	Year 5	Year 6	Year 7	Year 8	Year 9	Year 10
Total SWH Costs per Household ($)	954	1049	1154	1269	1396	1536	1689	1858	2044	2341
Total Subsidized Cost per Total number of Systems ($)	667	734	808	888	977	1075	1182	1301	1431	1639
Total Subsidized System Cost ($)	667450	-120903	0	-35223	-105456	-169515	-217462	-250190	-271369	-284500
Cumulative Subsidized Cost ($)	667450	546547	546547	511324	405867	236352	18891	-231299	-502669	-787168

Module 3: Electricity Savings

Input Parameters

RETScreen software [26] is used to estimate the electricity savings. The annual savings per system estimated by RETScreen [26] is $297 based on an electricity tariff of $0.069. Table **8.5** presents RETScreen [26] input parameters. The model calculates the annual electricity saved as a result of replacing traditional systems by SWH based on the electricity tariff without the subsidy provided by the government. It is assumed that the 90% of the electricity tariff is subsidy, as shown in Table **8.6**. The cumulative amount of electricity saved is $3,998,809 as shown in Table **8.6**.

Table 8.5. SWH Module 3: RETScreen [26] parameters.

Input Parameters	Value
Occupancy Rate	90%
Daily Hot Water Usage Estimate (L/day)	375
Hot Water Temperature °C	50
Operating Days per Week	7
Supply Water Minimum Temperature °C	18.7
Supply Water Maximum Temperature °C	23.6

(Table 8.5) contd.....

Input Parameters	Value
Gross Area per Solar Collector (m^2)	3.12
Aperture Area per Solar Collector (m^2)	1.89
Fr (tau alpha) Coefficient	0.41
Fr (UL) Coefficient	2.35
Number of Collectors	2
Capacity (kW)	1.32
Initial Cost (US$)	1000
Miscellaneous Losses (%)	3
Storage Capacity per Square Area (L/m^2)	151.2
Conventional Fuel Type	Electricity
Electricity Rate (US$)	0.069
Inflation Rate (%)	7.6
Project Life (Years)	10
Debt Ratio (% of Initial Cost)	0

Table 8.6. SWH – Lease: Module 3 - Electricity savings.

Parameters	Year 1	Year 2	Year 3	Year 4	Year 5	Year 6	Year 7	Year 8	Year 9	Year 10
Electricity Tariff ($/kWh)	0.069	0.069	0.069	0.069	0.069	0.069	0.069	0.069	0.069	0.069
Percentage of Electricity Tariff Subsidy (%)	90	90	90	90	90	90	90	90	90	90
Annual Electricity Savings per Collector ($)	297	326.7	359.37	395.31	434.84	478.32	526.15	578.77	636.65	700.31
Annual Total Electricity Savings ($)	297000	163350	76370	42003	23102	12706	6988	3844	2114	1163
Cumulative Electricity Saved (kWh)	297000	460350	536720	578723	601825	614531	621519	625363	627477	628639

(Table 8.6) contd.....

Parameters	Year 1	Year 2	Year 3	Year 4	Year 5	Year 6	Year 7	Year 8	Year 9	Year 10
Annual Total Subsidized Cost of Electricity Saved ($)	20493	11271	5270	2898	1594	877	482	265	146	80
Annual Total Electricity Saved ($)	204930	112712	52695	28982	15940	8767	4822	2652	1459	802
Cumulative Cost of Electricity Saved ($)	204930	317642	370337	399319	415259	424026	428848	431500	432959	433761
Cumulative Cost of Electricity Saved over the whole Duration ($)	204930	522572	892908	1292227	1707486	2131512	2560360	2991861	3424820	3858581

Module 4: Cash Flow

In the fourth module, shown in Table **8.7**, the model calculates the cash flow of the project over the 10 years. Then, based on the cash flow the ratio between the loan payback and the amount used to invest in the project using optimization. The objective, constraints and variables used in the optimization are as follows:

1. **Objective:** Maximize the profits.
2. **Variables:** Percentage of money used to reinvest in the following year over the whole project duration.
3. **Constraints:**

   a. The percentages of loan payback and reinvestment shall be less than 1 and greater than zero.

   b. The amount of money paid to payback the loan in the 10th year shall be zero.

   c. The total amount of money paid to payback the loan shall be greater than or equal to $1,650,000 and less than or equal to $3,000,000.

Table 8.7. SWH – Lease: Module 4 - Cash flow.

Parameters	Year 1	Year 2	Year 3	Year 4	Year 5	Year 6	Year 7	Year 8	Year 9	Year 10
Cash Flow ($)	976001	255702	265743	376451	541905	751815	1007522	1292467	1540152	1601342
Percentage Investment (%)	41	79	74	80	88	95	98	95	84	0.00
Percentage Line of Credit (%)	59	21	26	20	12	5	2	5	16	100
Money Used in Investment ($)	400450	201903	197888	300245	475451	712071	986518	1228429	1286570	0
Cumulative Money Used in Investment ($)	400450	602354	800241	1100486	1575936	2288008	3274526	4502955	5789524	5789524
Money Used for Line of Credit Payment ($)	575551	53798	67855	76206	66454	39744	21004	64038	253582	1601342
Cumulative Money Used for Line of Credit Payment ($)	575551	629349	697204	773410	839865	879608	900612	964651	1218233	2819575
Cumulative Cash Flow ($)	976001	1231703	1497446	1873896	2415801	3167616	4175138	5467606	7007757	8609099
Profits ($)	8609099									
Line of Credit Payment ($)	-1650000	-1181894	-1240905	-1290355	-1335564	-1396021	-1491905	-1617991	-1709348	-1601342
Left from Line of Credit Payment ($)	-1074449	-1128096	-1173050	-1214149	-1269110	-1356277	-1470901	-1553952	-1455765	-0.11
Line of Credit Payback Period (Years)	>10									
Net Present Value ($)	4,706,760									

Module 5: Results

The results of the model show that the loan payback period is 10 years while the total profits are estimated to be $8,609,099. Table **8.8** compares the percentages of the amount of reinvestment against the amount of loan payback over 10 years. Figs. (**8.17** to **8.20**) summarizes the main results of the current scenario. Fig. (**8.18**) shows that the consumers' annual savings is equal to their annual payments to the investor because all the savings are paid to the investor. Fig. (**8.19**), shows a comparison between the NPV in three cases. In the first case the annual profits are equally divided between loan payback payment and reinvestment in the following year while the second case the annual profits are totally paid to the loan payback. The third case uses optimization to obtain the loan/reinvestment ratio. The comparison indicates that the NPV of the optimized lease scenario is the most profitable.

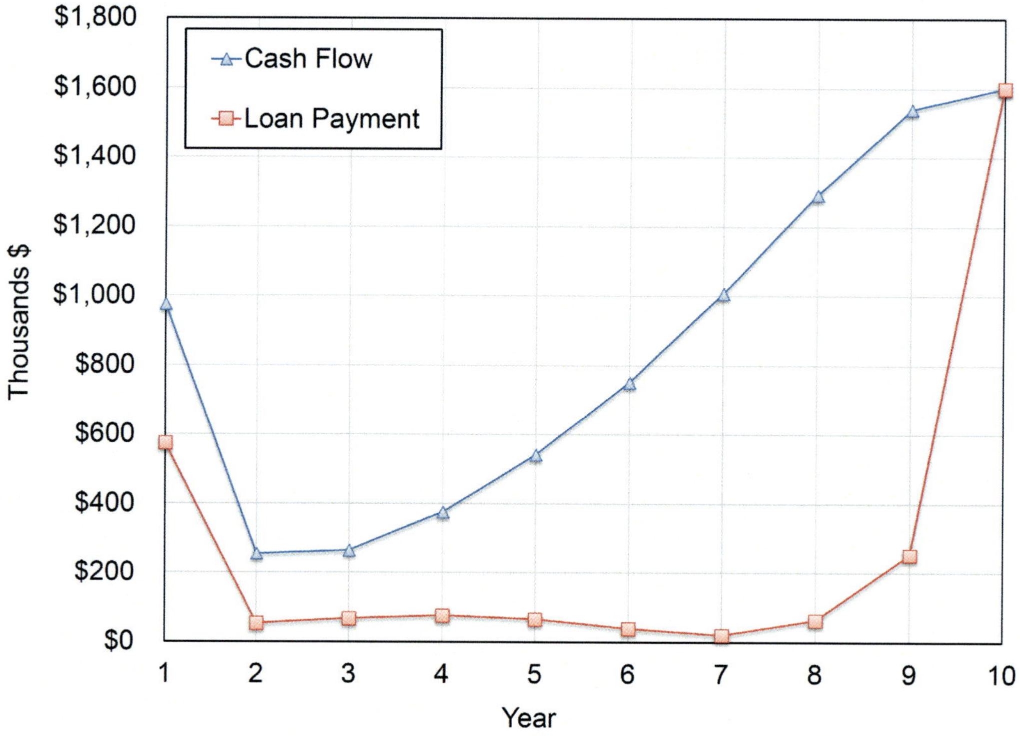

Fig. (8.17). SWH – Lease: Cash flow and loan payback.

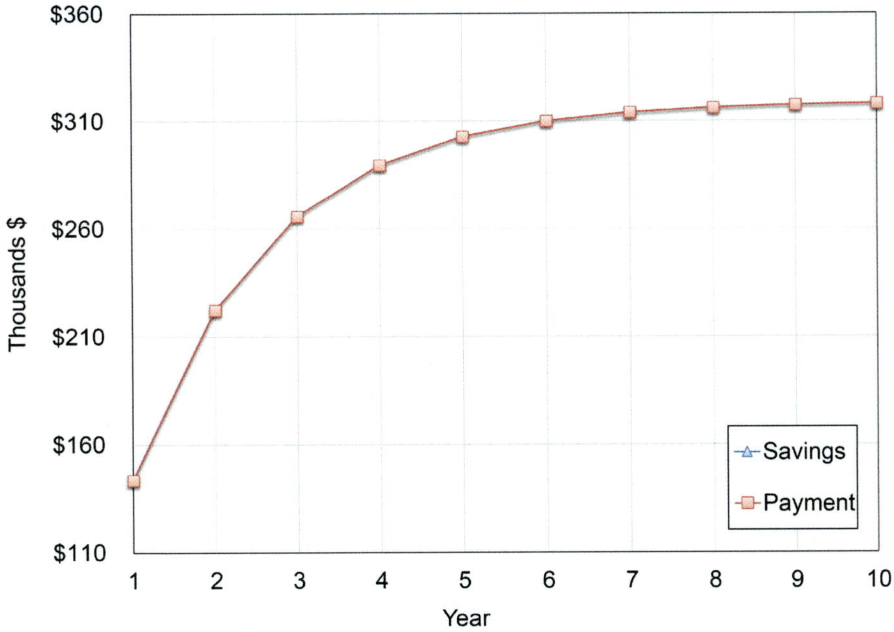

Fig. (8.18). SWH – Lease: Consumers' annual savings and payments.

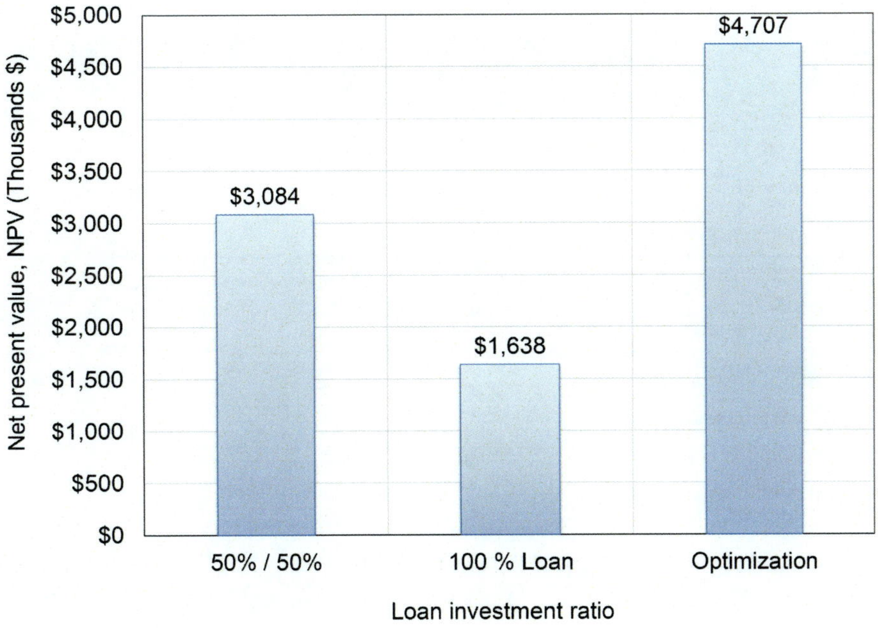

Fig. (8.19). SWH – Lease: Net present value.

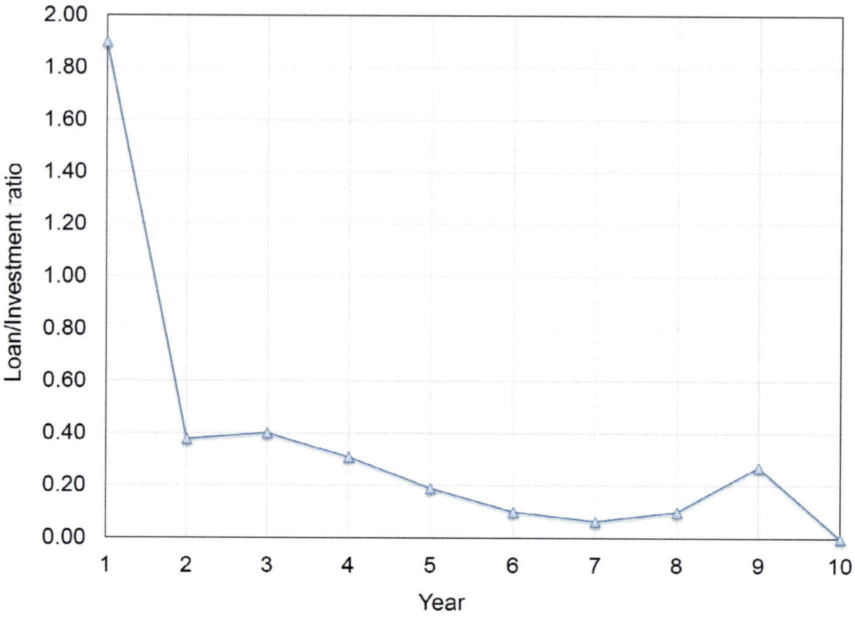

Fig. (8.20). Line of credit payback *vs.* reinvestment ratio (70%).

Table 8.8. SWH percentage of reinvestment & line of credit payback payment.

Amount of Loan Payback *Vs.* Amount of reinvestment										
	Yr 1	Yr 2	Yr 3	Yr 4	Yr 5	Yr 6	Yr 7	Yr 8	Yr 9	Yr 10
Amount of Reinvestment	0.41	0.79	0.74	0.80	0.88	0.95	0.98	0.95	0.84	0
Amount of Loan Payback	0.59	0.21	0.26	0.20	0.12	0.05	0.02	0.05	0.1	1

PV Model

Module 1: Project Data

Input Parameters

The input parameters for Module 1 are summarized in (Table **8.9**). The number of households that adopted the PV model in the first year is 204 which is 50% of the targeted population and also it is the maximum number of systems that can be purchased by the fund provided, as shown in Table **8.9**. In the following years the number of households that adopt the system decreases until it becomes 37 households in the 10th year. The reason behind this is that the model performs a check on the amount of money used to invest in the project in the previous year and verifies if it is sufficient to pay the subsidy for 50% of the targeted population in the current year or not. If that amount is not enough then the maximum number of subsidized systems that can be installed is shown instead. The cumulative number of systems installed over the total duration of the project is 400 households.

Table 8.9. SWH – Lease: Module 4 - Cash flow.

Fund Amount ($)	1500000									
Number of Households	2000									
Average Roof Area (m^2)	300									
Total Roof Area (m^2)	600000									
Percentage of Useful Area (%)	90									
Total Useful Area (m^2)	480000									
Loan Interest Rate (%)	10									
Subsidy Percentage (%)	70									
Percentage Adoption (%)	50									
Parameters	**Year 1**	**Year 2**	**Year 3**	**Year 4**	**Year 5**	**Year 6**	**Year 7**	**Year 8**	**Year 9**	**Year 10**
Annual Number of Installations	204	45	0	11	28	36	27	36	38	37
Annual Number of Installations Left	1796	1751	1751	1740	1712	1676	1649	1613	1575	1538
Cumulative Number of Installations	204	249	249	260	288	324	351	387	425	462
Average Roof Area (m^2)	300	300	300	300	300	300	300	300	300	300

(Table 8.9) contd.....

Percentage of Useful Area (%)	90	90	90	90	90	90	90	90	90	90
Annual Area Installed (m^2)	55102	12048	0	3041	7515	9679	7412	9673	10311	9997
Area Installed Left (m^2)	484898	472850	472850	469809	462293	452615	445203	435530	425220	415223
Cumulative Area Installed (m^2)	55102	67150	67150	70191	77707	87385	94797	104470	114780	124777

Module 2: Appliances and Energy Demand

Inputs Parameters

The average number of occupants per household specified in the first scenario is 5 occupants. Six types of household appliances are selected to be operated using PV modules instead of grid electricity namely; air conditioner, cooking range, plasma TV, incandescent bulb, microwave and refrigerator. The demand of the six appliances is calculated based on the number of appliances used as well as the number of hours each appliance is used as shown in Table **8.10** and Table **8.11**. The daily demand per household is 40.360 kWh.

Table 8.10. PV – Lease: Module 2 - Household appliances selection.

Appliance 1			Appliance 2			Appliance 3			Appliance 4			Appliance 5			Appliance 6			
Room Air Conditioner			Cooking Range			Plasma TV			Incandescent bulb			Microwave			Refrigerator			
Typical Wattage (W)	Number of Appliances	Hours Used (h)	Typical Wattage (W)	Number of Appliances	Hours Used (h)	Typical Wattage (W)	Number of Appliances	Hours Used (h)	Typical Wattage (W)	Number of Appliances	Hours Used (h)	Typical Wattage (W)	Number of Appliances	Hours Used (h)	Typical Wattage (W)	Number of Appliances	Hours Used (h)	Total Demand (Wh/ day/Household)
1000	2	5	1000	1	5	339	1	5	60	25	10	1500	1	5	188	1	8	40360

Module 3: PV System Data

Input Parameters

The type of PV system selected is Amorphous Silicon with a total cost of $10,500, as shown in Table **8.12**. A 10% increase in the systems costs due to inflation is taken into consideration for the following years. The total subsidized cost of the systems in the first year is $1,500,000. The amount of money used to reinvest in the project is deducted from the costs of the systems in the existing year, as shown in Table **8.12**.

Table 8.11. PV – Lease: Module 2 - Energy demand calculation.

Parameters	Year 1	Year 2	Year 3	Year 4	Year 5	Year 6	Year 7	Year 8	Year 9	Year 10
Average Number of Occupants per Household	5	5	5	5	5	5	5	5	5	5
Daily Average Selected Appliances Consumption (kWh/day/Household)	40.36	40.36	40.36	40.36	40.36	40.36	40.36	40.36	40.36	40.36
Total Annual Demand (kWh)	3006408	657345	0	165932	410043	528083	404390	527739	562559	545421

Table 8.12. PV – Lease: Module 3 - PV system type.

Parameters	Year 1	Year 2	Year 3	Year 4	Year 5	Year 6	Year 7	Year 8	Year 9	Year 10
PV System Type	Amorphous Silicon									
Initial Cost per (5 kW) System ($)	10500	11550	12705	13976	15373	16910	18601	20462	22508	24758
Initial Subsidized Cost per (5 kW) System ($)	7350	8085	8894	9783	10761	11837	13021	14323	15755	17331
Module Size (m2)	1.4	1.4	1.4	1.4	1.4	1.4	1.4	1.4	1.4	1.4
Number of Modules Installed	39359	4303	4339	4337	4299	4934	3859	4038	3897	3742
Annual Total Initial Cost of Systems ($)	2142857	257692	285864	314275	342670	432618	372218	428386	454821	480337
Total Subsidized Cost of PV Installed ($)	1500000	0	0	0	0	43262	0	0	0	0
Cumulative Subsidized Cost of PV Installed ($)	1500000	1500000	1500000	1500000	1500000	1543262	1543262	1543262	1543262	1543262
Total Energy Produced (kWh)	3006408	328672	331458	331273	328367	376874	294778	308419	297683	285803
Cumulative Energy Produced (kWh)	3006408	3335081	3666539	3997812	4326179	4703053	4997831	5306250	5603933	5889736

Module 4: Electricity Savings

Input Parameters

RETScreen software [26] is used to estimate the electricity savings. The annual savings per system estimated by RETScreen [26] is $2124 based on an electricity tariff of $0.069. Table **8.13** presents RETScreen [26] input parameters. The model calculates the annual electricity saved as a result of replacing traditional systems by PV based on the electricity tariff without the subsidy provided by the government. It is assumed that the 60% of the electricity tariff is subsidy, as shown in Table **8.13**. The cumulative amount of electricity saved is $7,685,626 as shown in Table **8.13**.

Table 8.13. PV – Lease: Module 4 - Electricity savings.

Parameters	Year 1	Year 2	Year 3	Year 4	Year 5	Year 6	Year 7	Year 8	Year 9	Year 10
Electricity Tariff (\$/kWh)	0.069	0.069	0.069	0.069	0.069	0.069	0.069	0.069	0.069	0.069
Annual Electricity Savings (kWh)	3006408	328672	331458	331273	328367	376874	294778	308419	297683	285803
Cumulative Electricity Savings (kWh)	3006408	3335081	3666539	3997812	4326179	4703053	4997831	5306250	5603933	5889736
Annual Total Cost of Electricity Saved with Subsidy (\$)	437755	52127	57826	63573	54338	87512	75294	86656	184008	97165
Annual Cumulative Cost of Electricity Saved with Subsidy (\$)	437755	489883	547709	611282	665620	753133	828427	915083	1099091	1196256
Percentage of Electricity Subsidized Cost (%)	0.60	0.60	0.60	0.60	0.60	0.60	0.60	0.60	0.60	0.60
Annual Total Cost of Electricity Saved without Subsidy (\$)	515384	56344	56821	56790	56291	64607	50533	52872	51031	48995
Cumulative Cost of Electricity Saved without Subsidy (\$)	515384	571728	628550	685339	741631	806238	856771	909643	960674	1009669
Cumulative Cost of Electricity Saved without Subsidy over the whole Duration(\$)	515384	1087112	1715662	2401001	3142632	3948869	4805640	5715283	6675957	7685626

Module 5: Cash Flow

In the fifth module, shown in Table **8.14**, the model calculates the cash flow of the project over the 10 years. Then, based on the cash flow the ratio between the loan payback and the amount used to invest in the project is determined using optimization. The objective, constraints and variables used in the optimization are as follows:

Table 8.14. PV – Lease: Module 5 - Cash flow.

Parameters	Year 1	Year 2	Year 3	Year 4	Year 5	Year 6	Year 7	Year 8	Year 9	Year 10
Cash Flow (\$)	360769	439650	439650	459562	508767	511518	620664	683993	751500	816950
Percentage Investment (%)	100	0.00	25	65	71	70	83	88	85	85
Percentage Line of Credit (%)	0.00	100	75	35	29	30	17	12	15	15
Money Used in Investment (\$)	360769	0	110192	299532	363716	357437	513111	601662	641666	694938
Cumulative Money Used in Investment (\$)	360769	360769	470961	770493	1134209	1491646	2004757	2606419	3248086	3943023

(Table 8.14) contd.....

Parameters	Year 1	Year 2	Year 3	Year 4	Year 5	Year 6	Year 7	Year 8	Year 9	Year 10
Money Used for Line of Credit Payment ($)	0	439650	329458	160030	145051	154081	107553	82331	109833	122013
Cumulative Money Used for Loan Payment ($)	0	439650	769109	929139	1074190	1228271	1335824	1418154	1527988	1650000
Cumulative Cash Flow ($)	360769	800419	1240070	1699632	2208399	2719917	3340581	4024574	4776074	5593024
Profits ($)	5593024									
Line of Credit Payment ($)	-1650000	-1418154	-978504	-649046	-489015	-343964	-189883	-82330	0	0
Left from Line of Credit Payment ($)	-1650000	-978504	-649046	-489015	-343964	-189883	-82330	0.18	0	0
Left from Line of Credit Payment ($) Profits	-1289231	-978504	-649046	-489015	-343964	-189883	-82330	0.18	109833	122013
Profit after Loan Payment and Investment	0	0	0	0	0	0	0	0.18	109833	122013
Total Amount of Profits ($)	694938									
Line of Credit Payback Period (Years)	8									
Net Present Value ($)	3211431									

1. **Objective:** Maximize the profits.
2. **Variables:** Percentage of money used to reinvest in the following year over the whole project duration.
3. **Constraints:**
   a. The percentages of loan payback and reinvestment shall be less than 1 and greater than zero.

   b. The amount of money paid to payback the loan in the 10th year shall be zero.

   c. The total amount of money paid to payback the loan shall be greater than or equal to $1,650,000 and less than or equal to $3,000,000.

Module 6: Results

The results of the model show that the loan payback period is 8 years while the total profits are estimated to be $5,593,024, as shown in (Table **8.14**). Table **8.15** compares the percentages of the amount of reinvestment against the amount of loan payback over 10 years. The table shows that the percentage determined for the loan is almost equal to the percentage used in the reinvestment. Figs. (**8.21** to **8.24**) summarizes the main results of the current scenario. Fig. (**8.22**) shows that the consumers' annual savings is equal to their annual payments to the investor because all the savings are paid to the investor. Fig. (**8.23**), shows a comparison between the NPV in three cases. In the first case, the annual profits are equally divided between loan payback payment and reinvestment in the following year. In the second case, the annual profits are totally paid to the loan payback. In the third case, optimization is used to obtain the loan/reinvestment ratio. The comparison indicates that the NPV of the optimized lease scenario is the most profitable.

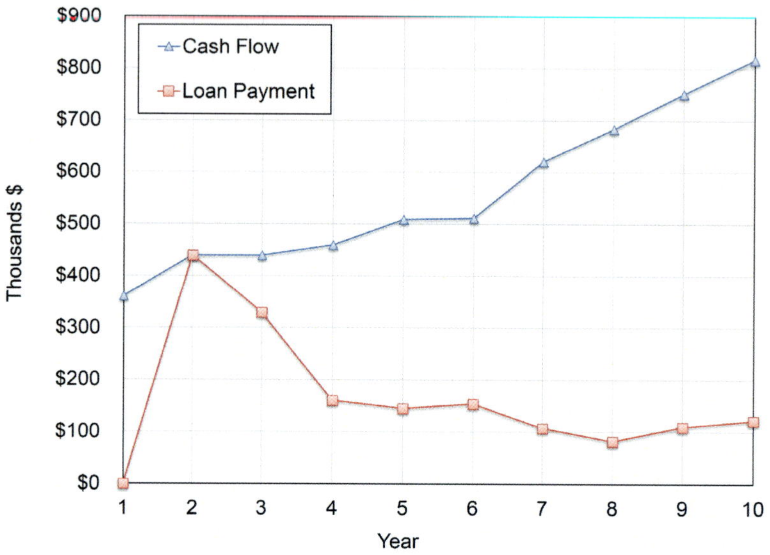

Fig. (8.21). PV – Lease: Cash flow and loan payback.

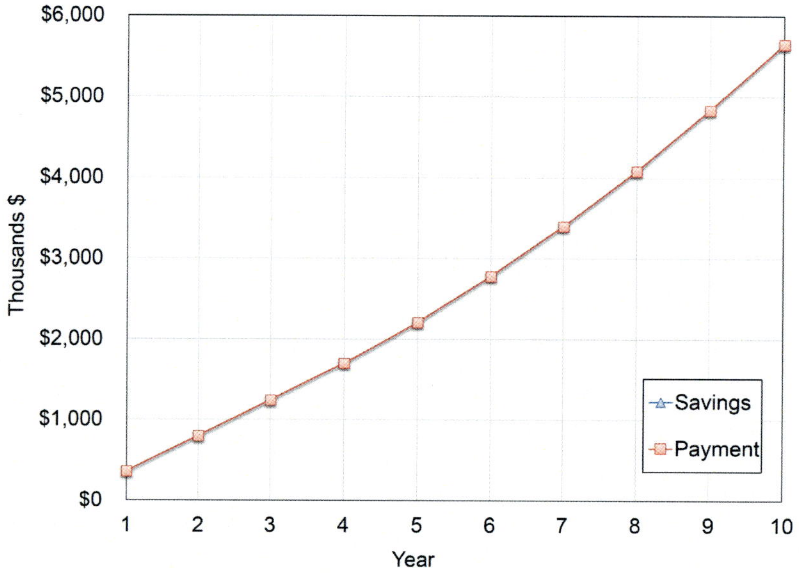

Fig. (8.22). PV – Lease: Consumers' annual savings and payments.

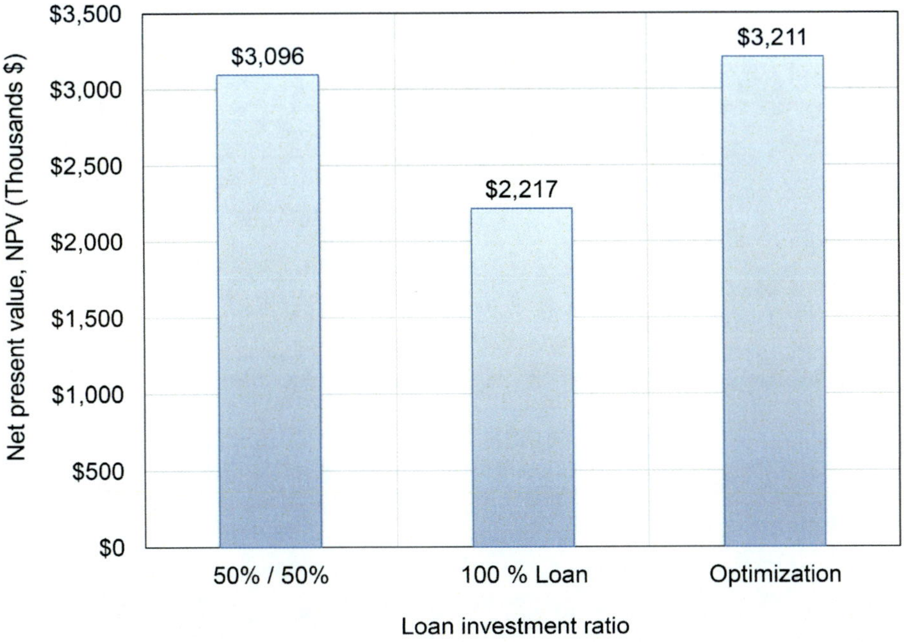

Fig. (8.23). PV – Lease: Net present value.

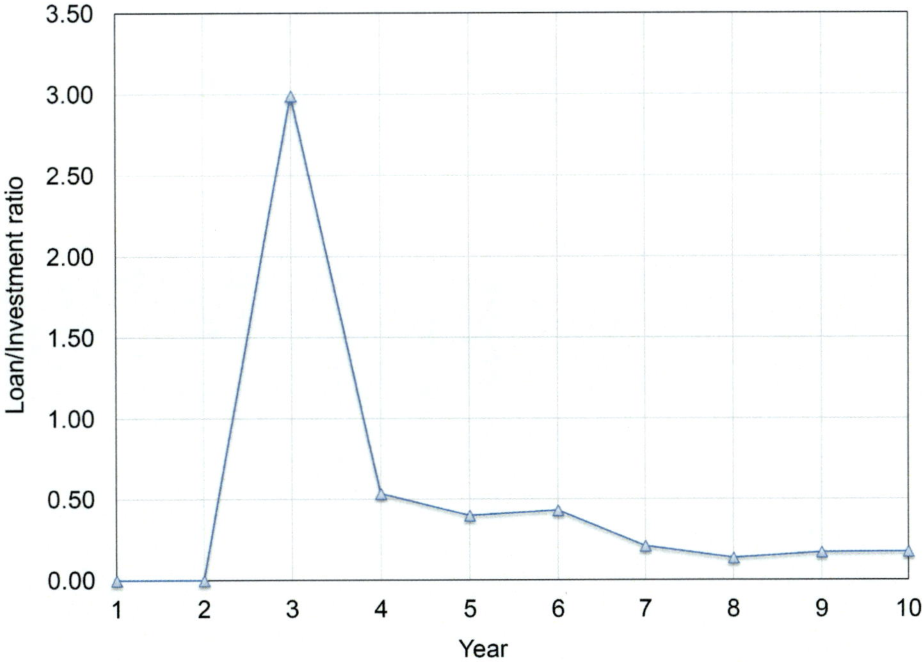

Fig. (8.24). Line of credit payback *vs.* reinvestment ratio.

Table 8.15. PV percentage of reinvestment & line of credit payback payment.

Ammount of Loan Payback *Vs.* Amount of reinvestment										
	Yr 1	Yr 2	Yr 3	Yr 4	Yr 5	Yr 6	Yr 7	Yr 8	Yr 9	Yr 10
Amount of Reinvestment	1	0	0.250	0.651	0.714	0.698	0.826	0.879	0.853	0.850
Amount of Loan Payback	0	1	0.749	0.348	0.850	0.301	0.173	0.120	0.146	0.149

Scenario 2: Loan

In this scenario, shown in Fig. (**8.25**), the investor funds the projects through a bank loan with an interest rate of 10%. The investor provides SWH & PV systems to the potential consumers with a subsidy of 70% of the systems' initial costs over a period of 10 years. The consumers install the systems and pay an annual fee to payback the cost of the system to the investor. The energy savings are estimated in order to determine the percentage of the fee that can be covered by the electricity savings. The model guides the investor on the amount of money paid to payback the line of credit and the amount used for the reinvestment of the project. The following sections show the inputs and the results of the scenario.

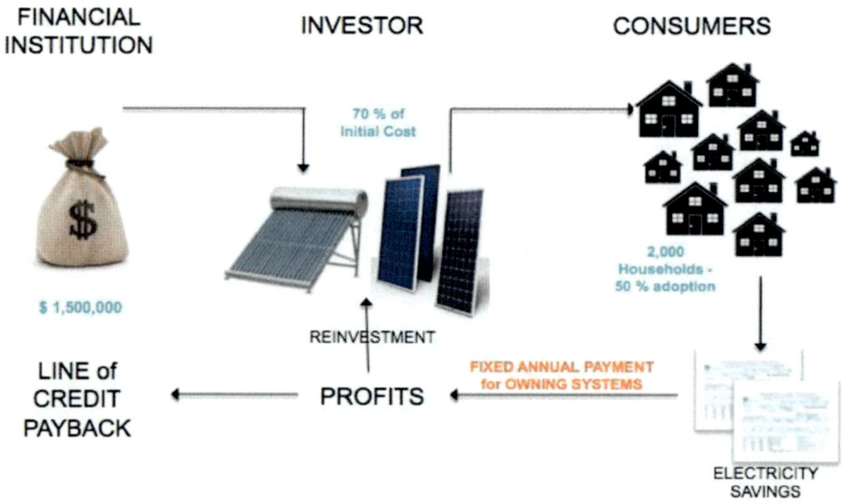

Fig. (8.25). Scenario 2: Loan.

SWH Model

Project data and SWH system data are the same as previously discussed in section 0. To calculate the electricity savings, it is assumed that the 50% of the electricity tariff is subsidy. The cumulative amount of electricity saved is $1,145,029 as shown in Table **8.16**. The cash flow for the current scenario is presented in Table **8.17**.

Table 8.16. SWH – Loan: Module 3 - Electricity savings.

Parameters	Year 1	Year 2	Year 3	Year 4	Year 5	Year 6	Year 7	Year 8	Year 9	Year 10
Electricity Tariff ($/kWh)	0.069	0.069	0.069	0.069	0.069	0.069	0.069	0.069	0.069	0.069
Percentage of Electricity Tariff Subsidy (%)	50	50	50	50	50	50	50	50	50	50
Annual Electricity Savings per Collector ($)	297	616	678	745	820	902	992	1091	1200	1320
Annual Total Electricity Savings ($)	297000	308000	169400	93170	51244	28184	15501	8526	4689	2579
Cumulative Electricity Saved (kWh)	297000	605000	774400	867570	918814	946997	962499	971024	975713	978292
Annual Total Subsidized Cost of Electricity Saved ($)	20493	21252	11689	6429	3536	1945	1070	588	324	178
Annual Total Electricity Saved ($)	40986	42504	23377	12857	7072	3889	2139	1177	647	356

(Table 8.16) contd.....

Parameters	Year 1	Year 2	Year 3	Year 4	Year 5	Year 6	Year 7	Year 8	Year 9	Year 10
Cumulative Cost of Electricity Saved ($)	40986	83490	106867	119725	126796	130686	132825	134001	134648	135004
Cumulative Cost of Electricity Saved over the whole Duration ($)	40986	124476	231343	351068	477864	608550	741375	875376	1010024	1145029
Annual Total Cost of Systems Installed ($)	89145	125474	135607	132588	122566	108401	91365	71961	50319	26383
Annual Cumulative Cost of Systems Installed ($)	89145	214619	350226	482814	605380	713781	805146	877107	927426	953809
Difference between Savings and Cost of System ($)	-48159	-131129	-243359	-363089	-478584	-583096	-672321	-743105	-792777	-818804
Annual Cost of Systems Paid by Users ($)	89145	214619	350226	482814	605380	713781	805146	877107	927426	953809

Module 5: Results

The results of the model show that the loan payback period is 9 years while the total profits are estimated to be $11,746,959, as shown in Table **8.17**. Table **8.18** compares the percentages of the amount of reinvestment against the amount of loan payback over 10 years. Figs. (**8.26** to **8.29**) summarizes the main results of the current scenario. Fig. (**8.27**) shows the consumers' annual savings and annual payments to the investor are not equal. As a result, the electricity savings in this scenario does not cover all the payments the consumer owes the investor.

Table 8.17. SWH – Loan: Module 4 - Cash flow.

Parameters	Year 1	Year 2	Year 3	Year 4	Year 5	Year 6	Year 7	Year 8	Year 9	Year 10
Cash Flow ($)	921695	482870	417540	597397	888036	1219440	1557028	1003881	1862201	2796871
Percentage Investment (%)	71	58	56	58	61	63	65	94	99	100
Percentage Line of Credit (%)	29	42	44	42	39	37	35	6	1	0
Money Used in Investment ($)	657101	281182	232211	347351	541241	771452	1014645	940696	1846318	2796871
Cumulative Money Used in Investment ($)	657101	938283	1170494	1517845	2059086	2830537	3845182	4785878	6632196	9429068
Money Used for Line of Credit Payment ($)	264594	201688	185330	250047	346795	447988	542383	63185	15883	0
Cumulative Money Used for Line of Credit Payment ($)	264594	466281	651611	901657	1248452	1696441	2238823	2302009	2317892	2317892
Cumulative Cash Flow ($)	921695	1404564	1822105	2419502	3307538	4526978	6084006	7087887	8950088	11746959
Profits ($)	11746959									
Line of Credit Payment ($)	-1650000	-1523947	-1454485	-1396071	-1260627	-1005216	-612950	-77624	-15882	0
Left from Line of Credit Payment ($)	-1385406	-1322259	-1269156	-1146025	-913832	-557227	-70567	-14438	0.60	0
Line of Credit Payback Period (Years)	9									
Net Present Value ($)	6,333,837									

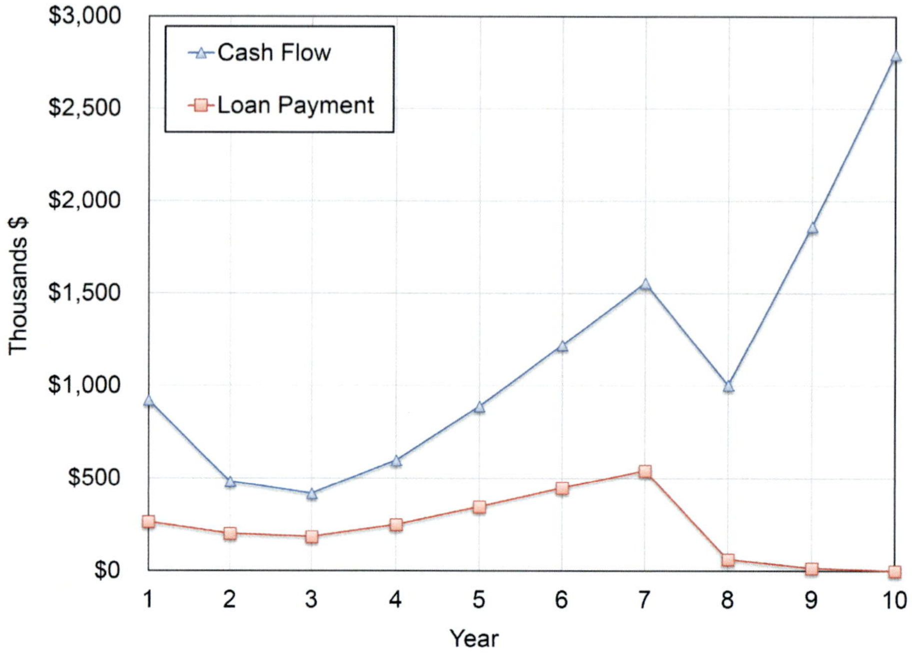

Fig. (8.26). SWH – Loan: Cash flow and loan payback.

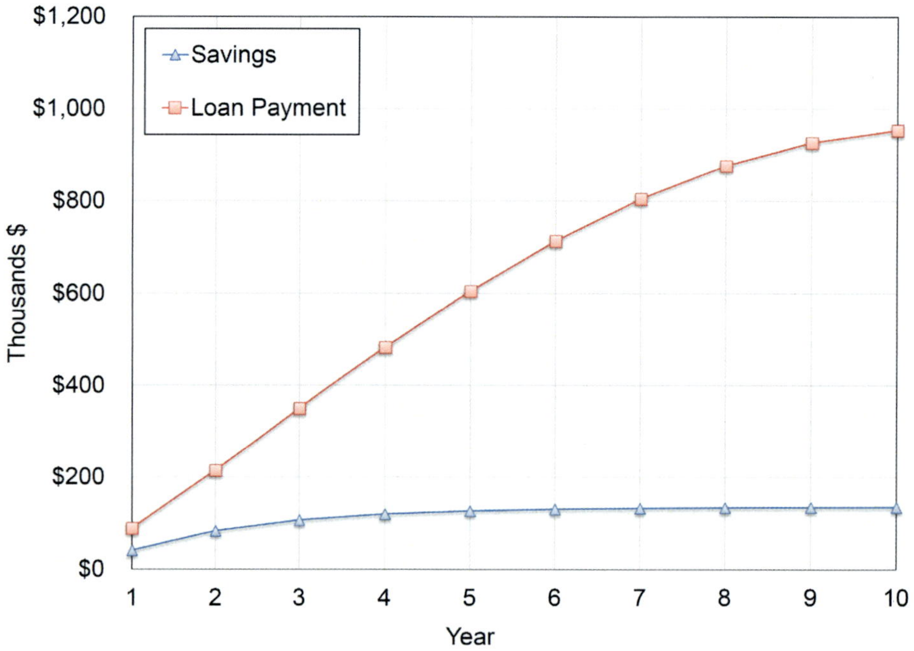

Fig. (8.27). SWH – Loan: Consumers' annual savings and payments.

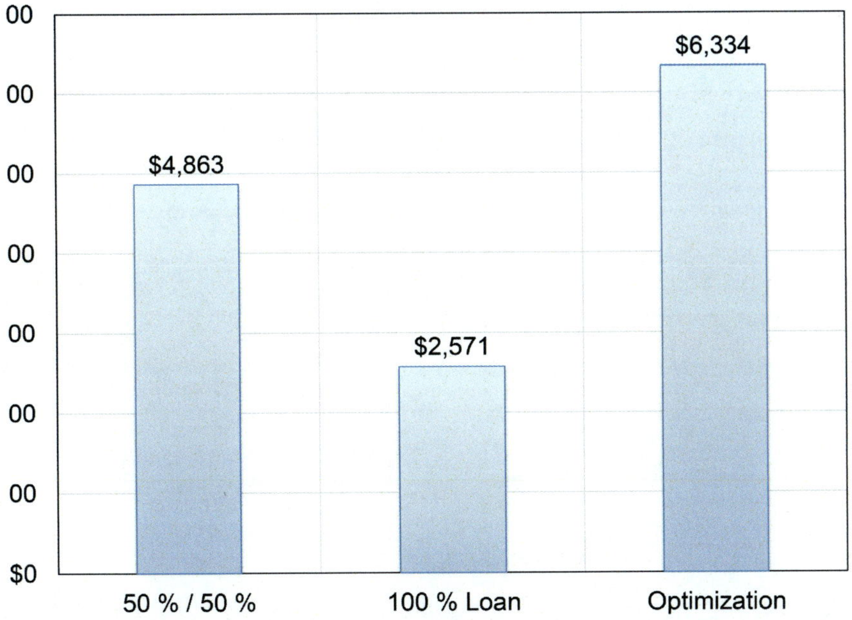

Fig. (8.28). SWH – Loan: Net present value.

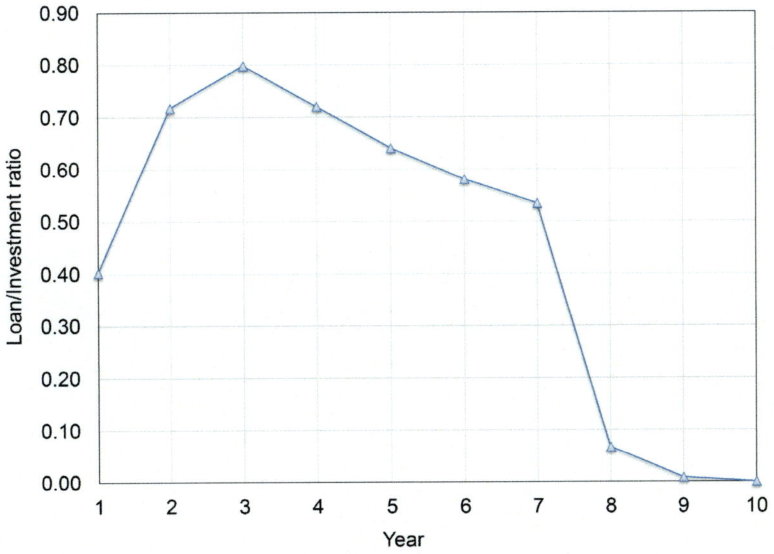

Fig. (8.29). SWH – Loan: Line of credit payback *vs.* reinvestment ratio.

PV Model

Project data and PV system data are the same as previously discussed in Section

0. To calculate the electricity savings, it is assumed that the 50% of the electricity tariff is subsidy, as shown in Table **8.19**. The cumulative amount of electricity saved without subsidy is $5,086,857, as shown in Table **8.19**. The cash flow for the current scenario is presented in Table **8.20**.

Table 8.18. SWH percentage of reinvestment & line of credit payback payment.

Ammount of Loan Payback *Vs.* Amount of reinvestment										
	Yr 1	Yr 2	Yr 3	Yr 4	Yr 5	Yr 6	Yr 7	Yr 8	Yr 9	Yr 10
Amount of Reinvestment	0.712	0.582	0.556	0.581	0.609	0.632	0.651	0.937	0.991	1
Amount of Loan Payback	0.287	0.417	0.443	0.418	0.390	0.367	0.348	0.062	0.008	0

Table 8.19. PV – Loan: Module 4 - Electricity savings.

Parameters	Year 1	Year 2	Year 3	Year 4	Year 5	Year 6	Year 7	Year 8	Year 9	Year 10
Electricity Tariff ($/kWh)	0.069	0.069	0.069	0.069	0.069	0.069	0.069	0.069	0.069	0.069
Annual Electricity Savings (kWh)	12626914	896915	1688203	2347977	2936139	4009622	2478515	1239257	619629	309814
Cumulative Electricity Savings (kWh)	12626914	13523830	15212033	17560010	20496149	24505771	26984285	28223543	28843171	29152986
Annual Total Cost of Electricity Saved with Subsidy ($)	2185714	144662	299516	458228	587913	946840	643809	354095	389505	107114
Annual Cumulative Cost of Electricity Saved with Subsidy ($)	2185714	2330376	2629892	3088120	3676033	4622873	5266682	5620778	6010282	6117396
Percentage of Electricity Subsidized Cost (%)	40	40	40	40	40	40	40	40	40	40

(Table 8.19) contd.....

Parameters	Year 1	Year 2	Year 3	Year 4	Year 5	Year 6	Year 7	Year 8	Year 9	Year 10
Annual Total Cost of Electricity Saved without Subsidy ($)	199837	13226	27384	41895	268447	86568	58863	32374	35612	9793
Cumulative Electricity Cost Saved without Subsidy ($)	199837	213063	240447	282342	550789	637358	696220	728595	764206	774000
Cumulative Electricity Cost Saved without Subsidy over the whole Duration ($)	199837	412900	653347	935689	1486479	2123836	2820056	3548651	4312857	5086857
Annual Payment ($)	239061	239977	250056	263411	275047	286232	263556	216263	154240	81660
Cumulative Annual Payment ($)	239061	479038	729094	992505	1267552	1553784	1817340	2033603	2187843	2269504

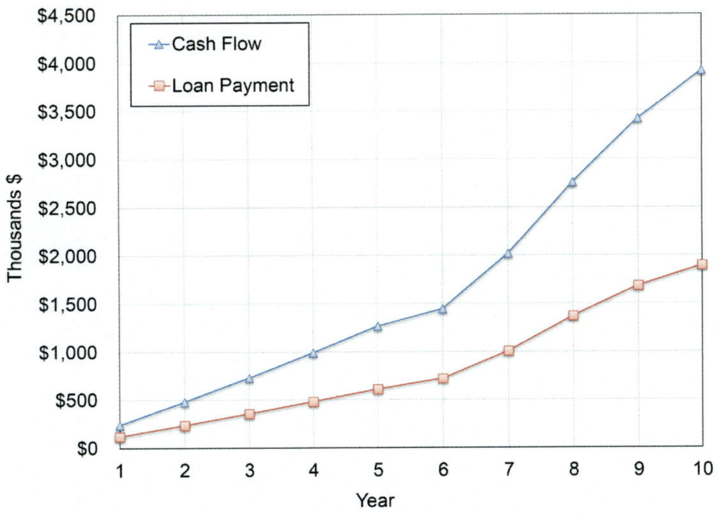

Fig. (8.30). PV – Lease: Cash flow and loan payback.

Module 6: Results

The results of the model show that the loan payback period is 5 years while the total profits are estimated to be $17,261,836, as shown in Table **8.20**. Table **8.21** compares the percentages of the amount of reinvestment against the amount of loan payback over 10 years. The table shows that the percentage determined for the loan is almost equal to the percentage used in the reinvestment. Figs. (**8.30** to **8.33**) summarizes the main results of the current scenario.

Table 8.20. PV – Loan: Module 5 - Cash flow.

Parameters	Year 1	Year 2	Year 3	Year 4	Year 5	Year 6	Year 7	Year 8	Year 9	Year 10
Cash Flow ($)	239061	479038	729094	992505	1267552	1444196	2018400	2758131	3417357	3916502
Percentage Investment (%)	49	51	51	51	52	50	50	50	51	52
Percentage Line of Credit (%)	51	49	49	49	48	50	50	50	49	48
Money Used in Investment ($)	117203	242663	371250	510671	657528	722664	1011411	1387299	1733780	2023142
Cumulative Money Used in Investment ($)	117203	359866	731116	1241787	1899315	2621979	3633390	5020689	6754469	8777611
Money Used for Line of Credit Payment ($)	121859	236375	357844	481834	610024	721532	1006989	1370832	1683576	1893360
Cumulative Money Used for Loan Payment ($)	121859	358233	716078	1197911	1807935	2529467	3536456	4907288	6590865	8484225
Cumulative Cash Flow ($)	239061	718100	1447193	2439698	3707250	5151446	7169846	9927977	13345334	17261836
Profits ($)	17261836									

(Table 8.20) contd.....

Parameters	Year 1	Year 2	Year 3	Year 4	Year 5	Year 6	Year 7	Year 8	Year 9	Year 10
Line of Credit Payment ($)	-1650000	-1686077	-1449702	-1091858	-610024	0	0	0	0	0
Left from Line of Credit Payment ($)	-1528141	-1449702	-1091858	-610024	0	0	0	0	0	0
Left from Line of credit Payment ($) Profits	-1532797	-1449702	-1091858	-610024	0	721532	1006989	1370832	1683576	1893360
Total Amount of Profits ($)	2023142									
Line of Credit Payback Period (Years)	5									
Net Present Value ($)	8,722,883									

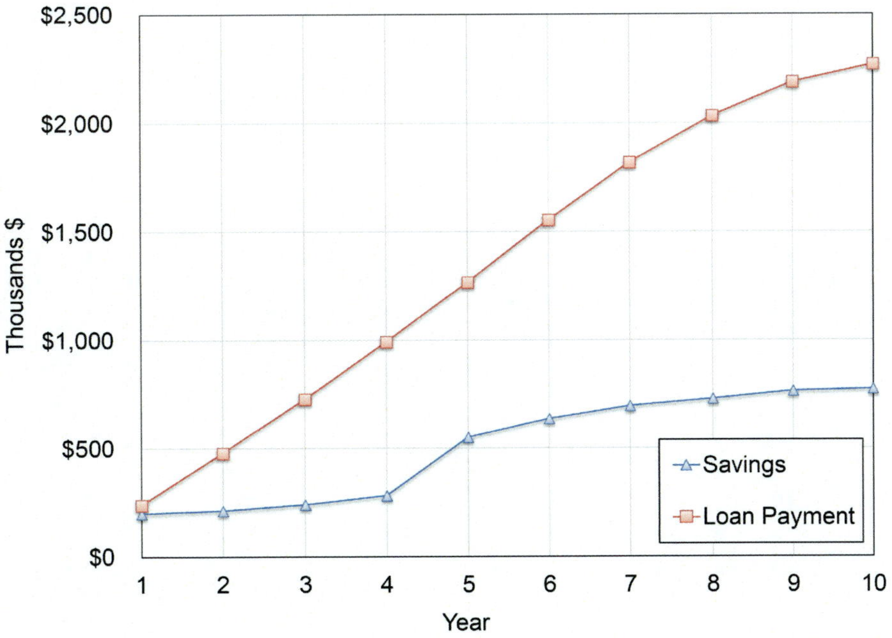

Fig. (8.31). PV – Lease: Consumers' annual savings and payments.

Table 8.21. PV percentage of reinvestment & line of credit payback payment.

Ammount of Loan Payback Vs. Amount of reinvestment										
	Yr 1	Yr 2	Yr 3	Yr 4	Yr 5	Yr 6	Yr 7	Yr 8	Yr 9	Yr 10
Amount of Reinvestment	0.490	0.506	0.509	0.514	0.518	0.500	0.501	0.503	0.507	0.516
Amount of Loan Payback	0.509	0.493	0.490	0.585	0.481	0.499	0.498	0.497	0.492	0.483

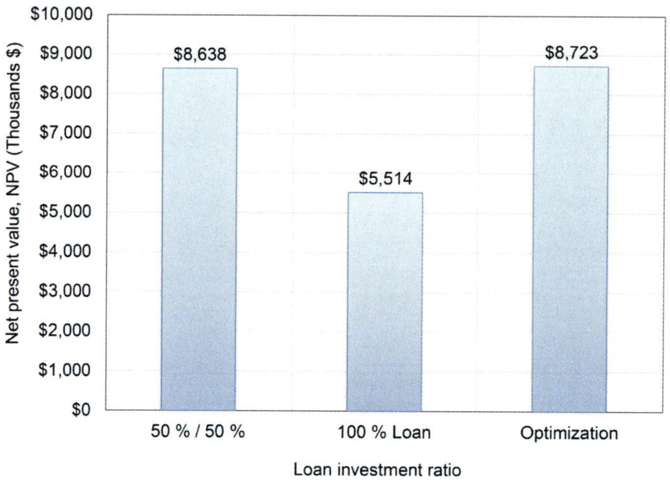

Fig. (8.32). Lease: Net present value.

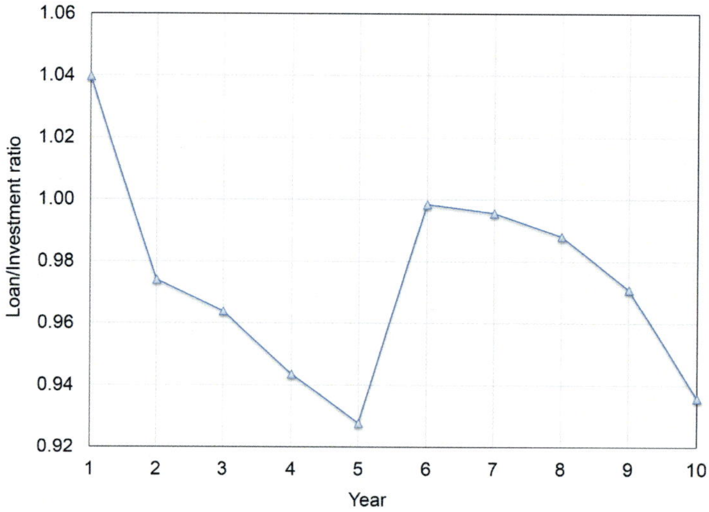

Fig. (8.33). Line of credit payback *vs.* reinvestment ratio.

Scenario 3: Power-Cut Tariff

Scenario 3, shown in Fig. (**8.34**), is similar to the preceding one, scenario 2. However, in the current scenario the consumers pay a higher rate additional fee if there is a power-cut. The following sections show each module's input parameters and the results of the first scenario.

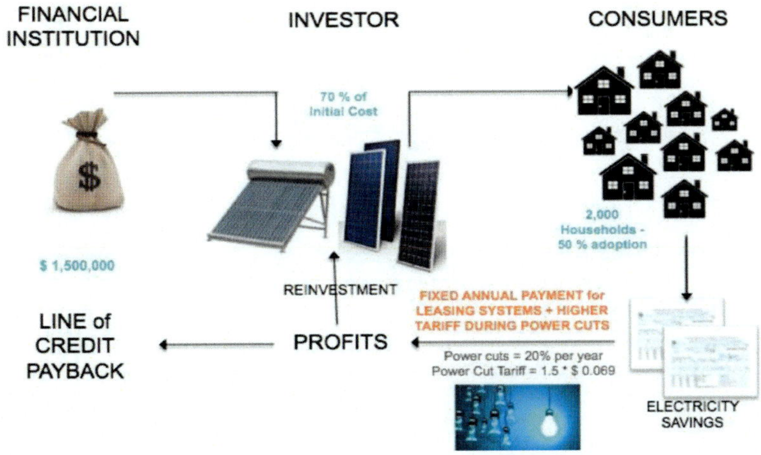

Fig. (8.34). Scenario 3: Power-cut tariff.

SWH Model

Project data and SWH system data are the same as previously discussed in section 0. To calculate the electricity savings, it is assumed that the 90% of the electricity tariff is subsidy, as shown in Table **8.22**. Also, it is assumed that the percentage of power cuts per year is 20%, the subsidized cost of electricity during power cuts is 150% of the electricity tariff and that 90% of the electricity tariff is subsidy. The cumulative amount of electricity saved is $3,141,344, as shown in (Table **8.22**). The cash flow for the current scenario is presented in (Table **8.23**).

Table 8.22. SWH – Power-Cut Tariff: Module 3 - Electricity savings.

Parameters	Year 1	Year 2	Year 3	Year 4	Year 5	Year 6	Year 7	Year 8	Year 9	Year 10
Electricity Tariff ($/kWh)	0.069	0.069	0.069	0.069	0.069	0.069	0.069	0.069	0.069	0.069

(Table 8.22) contd.....

Parameters	Year 1	Year 2	Year 3	Year 4	Year 5	Year 6	Year 7	Year 8	Year 9	Year 10
Annual Percentage of Power Cuts (%)	20	20	20	20	20	20	20	20	20	20
Annual Electricity Savings per Collector ($)	297	327	359	395	435	478	526	579	637	700
Percentage of Electricity Subsidized Cost (%)	90	90	90	90	90	90	90	90	90	90
Percentage of Electricity Subsidized Cost During Power Cuts (%)	150	150	150	150	150	150	150	150	150	150
Annual Total Electricity Saved ($)	297000	163350	82912	45602	25081	13795	7587	4173	2295	1262
Cumulative Electricity Saved ($)	297000	460350	543262	588864	613945	627740	635327	639500	641795	643057
Annual Total Subsidized Cost of Electricity Saved ($)	20493	11271	5721	3147	1731	952	524	288	158	87
Annual Total Cost of Electricity Saved ($)	163944	90169	45768	25172	13845	7615	4188	2303	1267	697
Cumulative Cost of Electricity Saved ($)	163944	254113	299881	325053	338898	346512	350700	353004	354271	354968

(Table 8.22) contd.....

Parameters	Year 1	Year 2	Year 3	Year 4	Year 5	Year 6	Year 7	Year 8	Year 9	Year 10
Cumulative Cost of Electricity Saved over the whole Duration ($)	163944	418057	717938	1042991	1381889	1728402	2079102	2432106	2786377	3141344
Annual Total Cost of Electricity Saved during Power Cuts ($)	61479	33813	17163	9440	5192	2855	1571	864	475	261
Cumulative Cost of Electricity Saved during Power Cuts ($)	61479	95292	112455	121895	104768	104768	104768	104768	104768	104768
Cumulative Cost of Electricity Saved during Power Cuts over the whole Duration ($)	61479	156771	84790	43765	24071	13239	7281	4005	2203	1211
Annual Total Cost of Electricity Saved ($)	225423	123983	62931	34612	19036	10470	5759	3167	1742	958
Cumulative Cost of Electricity Saved over the whole Duration ($)	225423	349406	412336	446948	465985	476455	482213	485380	487122	488080

Module 5: Results

The results of the model show that the loan payback period is 10 years while the total profits are estimated to be $8,454,398, as shown in (Table **8.23**). Table **8.24** compares the percentages of the amount of reinvestment against the amount of loan payback over 10 years. Figs. (**8.35** to **8.38**) summarizes the main results of the current scenario.

Table 8.23. SWH – Power-Cut Tariff: Module 4 - Cash flow.

Parameters	Year 1	Year 2	Year 3	Year 4	Year 5	Year 6	Year 7	Year 8	Year 9	Year 10
Cash Flow ($)	990346	262533	292328	428523	623880	874218	1182004	1171842	1286201	1342861
Percentage Investment (%)	39	77	76	82	91	98	100	80	77	0.00
Percentage Line of Credit (%)	61	23	24	18	9	2	0.00	20	23	100
Money Used in Investment ($)	385047	201914	221000	351946	566858	855160	1182004	942747	996291	0
Cumulative Money Used in Investment ($)	385047	586961	807961	1159907	1726765	2581924	3763928	4706675	5702966	5702966
Money Used for Line of Credit Payment ($)	605300	60619	71328	76577	57023	19058	0	229095	289910	1342861
Cumulative Money Used for Line of Credit Payment ($)	605300	665918	737246	813824	870846	889904	889904	1118999	1408909	2751770
Cumulative Cash Flow ($)	990346	1252879	1545207	1973730	2597611	3471828	4653832	5825674	7111875	8454736
Profits ($)	8454736									

(Table 8.23) contd.....

Parameters	Year 1	Year 2	Year 3	Year 4	Year 5	Year 6	Year 7	Year 8	Year 9	Year 10
Line of Credit Payment ($)	-1650000	-1149171	-1197407	-1238687	-1278321	-1343428	-1456807	-1602487	-1510731	-1342904
Left from Line of Credit Payment ($)	-1044700	-1088552	-1126079	-1162110	-1221298	-1324370	-1456807	-1373392	-1220822	-43
Line of Credit Payback Period (Years)	>10									
Net Present Value ($)	4,726,889									

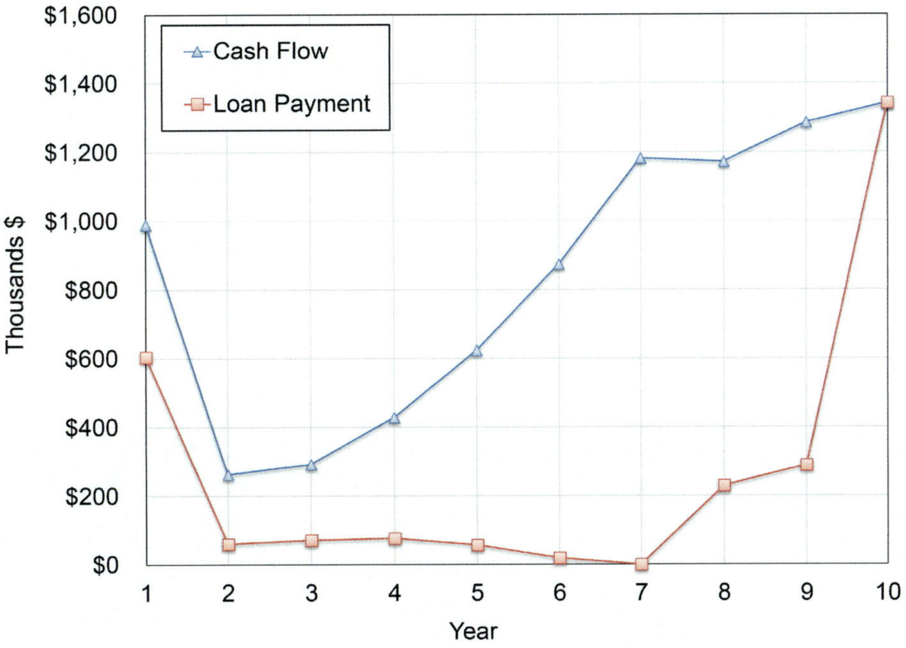

Fig. (8.35). SWH – Power-Cut Tariff: Cash flow and loan payback.

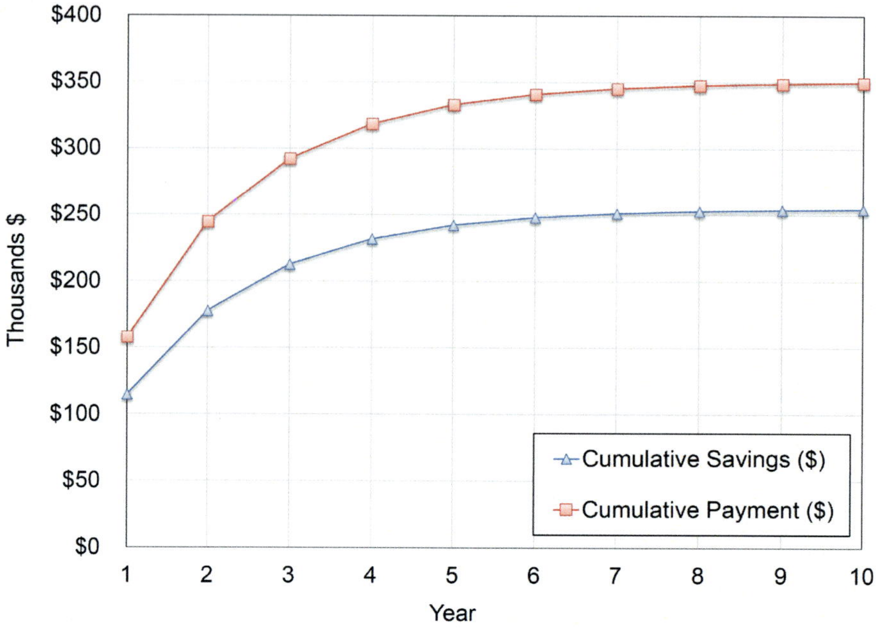

Fig. (8.36). SWH – Power-Cut Tariff: Consumers' annual savings and payments.

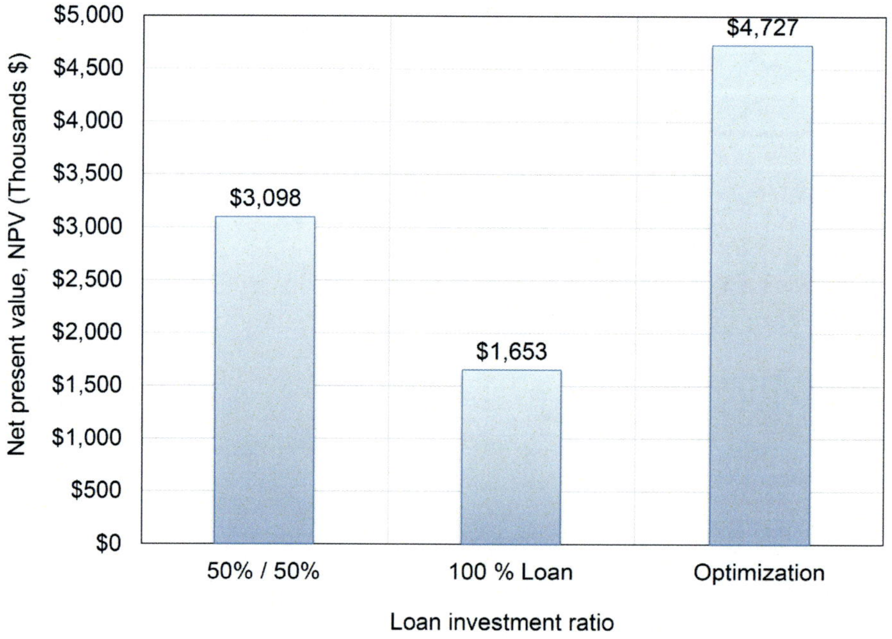

Fig. (8.37). SWH – Power-Cut Tariff: Net present value.

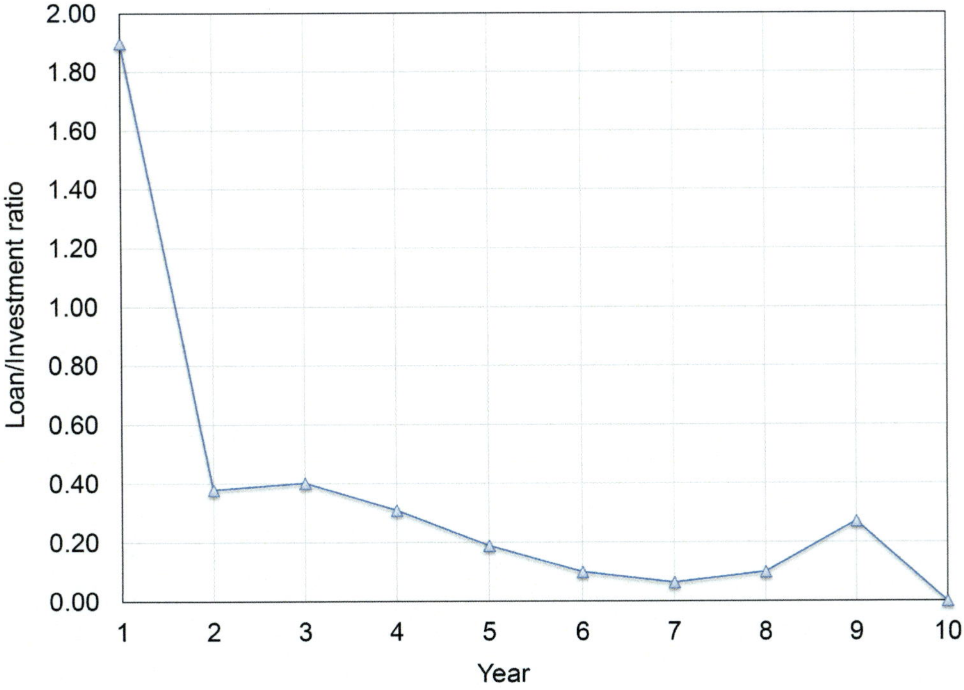

Fig. (8.38). SWH – Power-Cut Tariff: Line of credit payback *vs.* reinvestment ratio.

Table 8.24. SWH percentage of reinvestment & line of credit payback payment.

Ammount of Loan Payback *Vs.* Amount of reinvestment										
	Yr 1	Yr 2	Yr 3	Yr 4	Yr 5	Yr 6	Yr 7	Yr 8	Yr 9	Yr 10
Amount of Reinvestment	0.388	0.578	0.589	0.613	0.638	0.669	0.715	0.795	0.851	1
Amount of Loan Payback	0.884	0.422	0.410	0.386	0.361	0.331	0.284	0.204	0.149	0

PV Model

Project data and PV system data are the same as previously discussed in section 0. To calculate the electricity savings, it is assumed that the 40% of the electricity tariff is subsidy, as shown in (Table **8.25**). The cumulative amount of electricity saved is $6,454,492, as shown in Table **8.25**. The cash flow for the current scenario is presented in Table **8.26**.

Table 8.25. PV – Power-Cut Tariff: Module 4 - Electricity savings.

Parameters	Year 1	Year 2	Year 3	Year 4	Year 5	Year 6	Year 7	Year 8	Year 9	Year 10
Electricity Tariff ($/kWh)	0.069	0.069	0.069	0.069	0.069	0.069	0.069	0.069	0.069	0.069
Annual Percentage of Power Cuts (%)	20	20	20	20	20	20	20	20	20	20
Annual Electricity Savings (kWh)	3064785	644696	721703	807908	904410	740452	914879	1067868	1195422	1589676
Cumulative Electricity Savings (kWh)	3064785	3709481	4431184	5239092	6143502	6883954	7798833	8866701	10062123	11651799
Annual Total Cost of Electricity Saved with Subsidy ($)	540915	125163	154125	189788	130685	210470	286055	367279	904528	661564
Annual Cumulative Cost of Electricity Saved with Subsidy ($)	540915	666079	820204	1009992	1140677	1351147	1637202	2004480	2909008	3570573
Percentage of Electricity Subsidized Cost (%)	40	40	40	40	40	40	40	40	40	40
Percentage of Subsidy during Power Cuts	150	150	150	150	150	150	150	150	150	150
Annual Total Cost of Electricity Saved without Subsidy ($)	721221	166884	205500	253051	82689	280626	381406	489705	1206037	882086

(Table 8.25) contd.....

Parameters	Year 1	Year 2	Year 3	Year 4	Year 5	Year 6	Year 7	Year 8	Year 9	Year 10
Cumulative Electricity Cost Saved without Subsidy ($)	721221	888105	1093605	1346656	1429345	1709971	2091377	2581082	3787120	4669205
Cumulative Electricity Cost Saved without Subsidy over the whole duration ($)	721221	1609325	2702930	4049586	5478931	7188902	9280280	11861362	15648482	20317687
Annual Total Cost of Electricity Saved without Subsidy during Power Cuts ($)	270458	62582	77062	94894	65343	105235	143027	183639	452264	330782
Cumulative Electricity Cost Saved without Subsidy during Power Cuts ($)	270458	333039	410102	504996	570339	675573	818601	1002240	1454504	1785286
Total Cost of Electricity Saved over the whole duration ($)	991678	1221144	1503707	1851652	1999683	2385545	2909978	3583323	5241624	6454492

Module 6: Results

The results of the model show that the loan payback period is 5 years while the total profits are estimated to be $19,181,935, as shown in (Table **8.26**). Table **8.27** compares the percentages of the amount of reinvestment against the amount of loan payback over 10 years. The table shows that the percentage determined for the loan is almost equal to the percentage used in the reinvestment. Figs. (**8.39** to **8.42**) summarizes the main results of the current scenario.

Table 8.26. PV – Power-Cut Tariff: Module 5 - Cash flow.

Parameters	Year 1	Year 2	Year 3	Year 4	Year 5	Year 6	Year 7	Year 8	Year 9	Year 10
Cash Flow ($)	694175	854801	1052595	1296156	1000541	1586503	2036985	2508326	3669137	4518144
Percentage Investment (%)	45%	51%	53%	55%	50%	50%	50%	50%	50%	50%
Percentage Line of Credit (%)	55%	49%	47%	45%	50%	50%	50%	50%	50%	50%
Money Used in Investment ($)	309394	430653	549220	710341	500111	792536	1018391	1253788	1841241	2274332
Cumulative Money Used in Investment ($)	309394	740046	1289267	1999608	2499718	3292255	4310646	5564434	7405675	9680007
Money Used for Line of Credit Payment ($)	384781	406704	487436	580484	502115	794123	1018391	1254289	1827300	2246304
Cumulative Money Used for Line of Credit Payment ($)	384781	791485	1278921	1859405	2361520	3155643	4174034	5428324	7255624	9501928
Cumulative Cash Flow ($)	694175	1531532	2568187	3859013	4861239	6447897	8484680	10992758	14661299	19181935
Profits ($)	19181935									
Line of Credit Payment ($)	-1650000	-1474667	-1067963	-580527	-43	0	0	0	0	0
Left from Line of Credit Payment ($)	-1265219	-1067963	-580527	-43	502072	0	0	0	0	0

(Table 8.26) contd.....

Parameters	Year 1	Year 2	Year 3	Year 4	Year 5	Year 6	Year 7	Year 8	Year 9	Year 10
Left from Line of credit Payment Profits($)	-1340606	-1067963	-580527	-43	502072	794123	1018391	1254289	1827300	2246304
Line of Credit Payback Period (Years)	5									
Annual Payment per Consumer ($)	3337	3389	3497	3650	3326	3544	3820	4141	5341	5685
Net Present Value ($)	10,015,484									

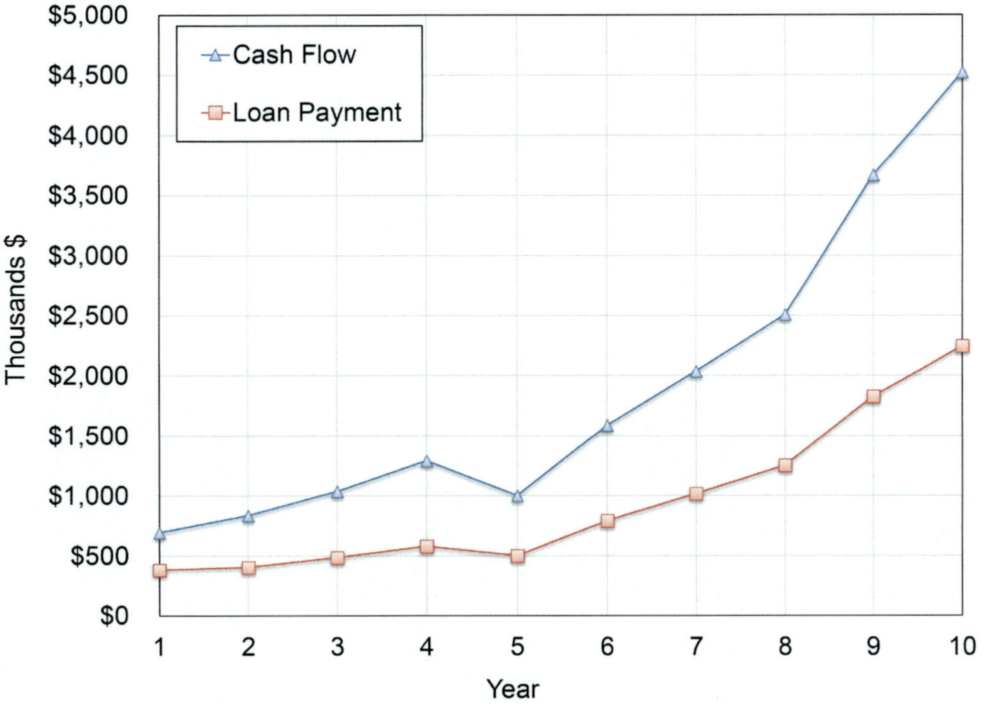

Fig. (8.39). PV – Power-Cut Tariff: Cash flow and loan payback.

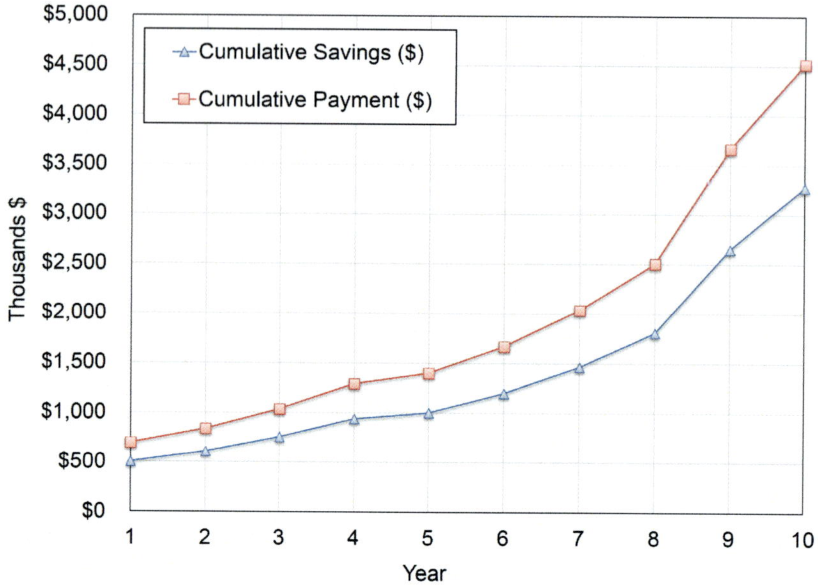

Fig. (8.40). PV – Power-Cut Tariff: Consumers' annual savings and payments.

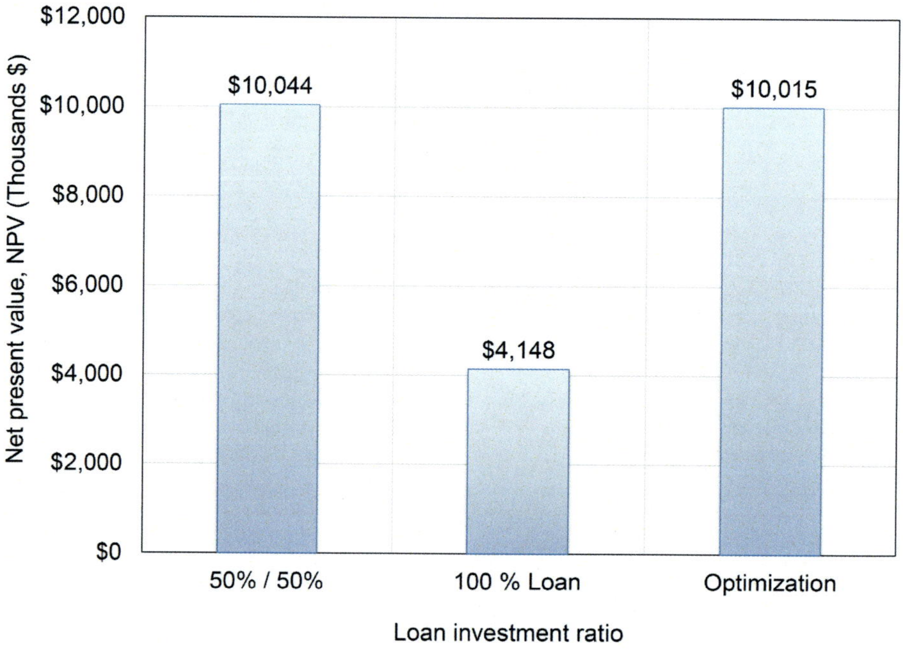

Fig. (8.41). PV – Power-Cut Tariff: Net present value.

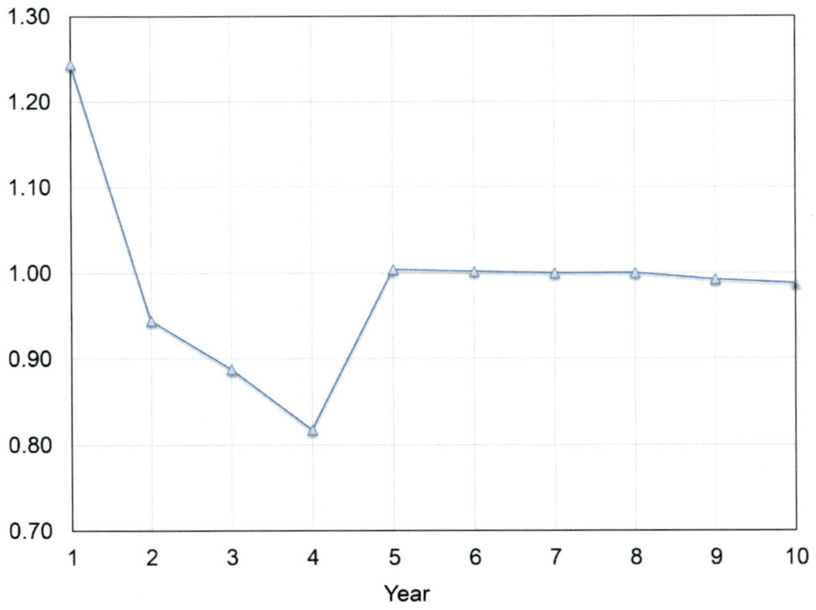

Fig. (8.42). PV – Power-Cut Tariff: Line of credit payback *vs.* reinvestment ratio.

Table 8.27. PV percentage of reinvestment & line of credit payback payment.

Ammount of Loan Payback Vs. Amount of reinvestment										
	Yr 1	Yr 2	Yr 3	Yr 4	Yr 5	Yr 6	Yr 7	Yr 8	Yr 9	Yr 10
Amount of Reinvestment	0.438	0.510	0.520	0.529	0.540	0.543	0.554	0.502	0.502	0.503
Amount of Loan Payback	0.561	0.489	0.479	0.470	0.459	0.456	0.445	0.497	0.497	0.496

DISCUSSION

The main target of this chapter is to examine the feasibility of adopting systems that use solar energy to replace traditional systems in residential urban Egyptian communities, hence reduce the amount of carbon dioxide emissions and conserve natural resources. The research methodology consisted of a literature review on the barriers that hinder the deployment of solar technologies, implementation strategies and initiatives in Egypt and other countries followed by developing a cost benefit analysis model that consists of two modules namely; solar water heating and photovoltaic modules and finally the model is exploited to perform different case studies. The cost benefit analysis model is developed in order to study the economic feasibility of the solar water heating and photovoltaic technologies. The model calculates the two systems' cash flows and the difference between the costs of the systems and the total amount of electricity that could be saved as a result of using the solar technologies. The cost of the system includes (i) the initial costs, (ii) operation and maintenance costs, and (iii) customs and taxes if applicable. The electricity savings are estimated using RETScreen software based on Cairo weather data including the solar irradiation. The model calculates the net present worth of each module and the percentage of money paid to payback the loan used to fund the project in terms of line of credit and the percentage of money used to reinvest in the project over a duration of 10 years.

The solar technologies economic barriers and implementation strategies in Egypt and worldwide are concluded from the literature review. Accordingly, seven scenarios are proposed from an investor and the government point of view to be applied by the model. The scenarios are; Loan, Lease, Electricity power cuts tariff, Feed in Tariff, Energy bonds, Tax reduction and subsidized systems with fixed payback rates.

A case study that incorporates three scenarios namely; Loan, Lease and Power-Cut Tariff is conducted. The scenarios selected for the case study are all from the perspective of the investor. The case study is located in one of the residential compounds, in New-Cairo in Cairo city in Egypt, with an initial fund of $1,500,000 for each module in terms of a line of credit to be paid back at an interest rate of 10% over a duration of 10 years. The targeted number of households is chosen to be 2000 household with an average number of occupants of 5 per household and a 90% occupancy rate. The basic scenario subsidy percentage is selected to be 70% of the units' initial costs while the adoption percentage is 50% of the potential adopters.

The results showed that the SWH module net present worth for the Lease, Loan and Power-Cut Tariff are $4,706,760, $6,333,837 and $4,726,889, respectively. On the other hand, the PV module net present worth for the Lease, Loan and Power Power-Cut Tariff are $3,211,431, $8,722,883 and $10,015,484, respectively. The SWH line of credit payback period for the Lease, Loan and Power-Cut Tariff are 10, 9, and 10 years while the PV line of credit payback period for the Lease, Loan and Power-Cut Tariff are 8, 5, and 5 years.

QUESTIONS

1. What are the geographical and seasonal limitations of solar energy?
2. What are the most significant technological challenges associated with solar energy technologies?
3. What are the costs of implementing solar energy technologies?
4. What are the political challenges of implementing renewable energy technology? Why are governments hesitant to implement renewable energy technologies?
5. What are the social and cultural challenges for the implementation of solar energy technologies? What are some reasons that a community might reject a particular technology?
6. Why is it important to begin implementing solar energy technologies?

NOTES

[2]Solar Water Heater

[3]New Taiwan dolla

[4]Interview with Representatives from: Sino Tech Inc., Khallouf Co., African Egyptian Group, City Pulse. December 20, 2010.

REFERENCES

[1] G.R. Timilsina, L. Kurdgelashvili, and P.A. Narbel, "Solar Energy: Markets, Economics and Policies", *Renew. Sustain. Energy Rev.,* vol. 16, no. 1, pp. 449-465, 2012. [http://dx.doi.org/10.1016/j.rser.2011.08.009]

[2] R.B. Chedid, "Policy Development for Solar Water Heaters: the Case of Lebanon", *Energy Convers. Manage.,* vol. 43, no. 1, pp. 77-86, 2002. [http://dx.doi.org/10.1016/S0196-8904(01)00008-5]

[3] G.R. Timilsina, T. Lefevre, and S. Shrestha, "Financing Solar Thermal Technologies under DSM Programs: An Innovative Approach to Promote Renewable Energy", *Int. J. Energy Res.,* vol. 24, no. 6, pp. 503-510, 2000. [http://dx.doi.org/10.1002/(SICI)1099-114X(200005)24:6<503::AID-ER599>3.0.CO;2-R]

[4] P. Veeraboina, and G.Y. Ratnam, "Analysis of the Opportunities and Challenges of Solar Water Heating System (SWHS) in India: Estimates from the Energy Audit Surveys & Review", *Renew. Sustain. Energy Rev.,* vol. 16, no. 1, pp. 668-676, 2012.

[http://dx.doi.org/10.1016/j.rser.2011.08.032]

[5] B.F. Mills, and J. Schleich, "Profits or Preferences? Assessing the Adoption of Residential Solar Thermal Technologies", *Energy Policy,* vol. 37, no. 10, pp. 4145-4154, 2009. [http://dx.doi.org/10.1016/j.enpol.2009.05.014]

[6] A. Faiers, and C. Neame, "Consumer Attitudes Towards Domestic Solar Power Systems", *Energy Policy,* vol. 34, no. 14, pp. 1797-1806, 2006. [http://dx.doi.org/10.1016/j.enpol.2005.01.001]

[7] E. Ndzibah, "“Diffusion of Solar Technology in Developing Countries: Focus Group Study in Ghana,” Manag. Environ. Qual", *Int. J.,* vol. 21, no. 6, pp. 773-784, 2010.

[8] P. Zhai, "Analyzing Solar Energy Policies Using a Three-Tier Model: A Case Study of Photovoltaics Adoption in Arizona, United States", *Renew. Energy,* vol. 57, pp. 317-322, 2013. [http://dx.doi.org/10.1016/j.renene.2013.01.058]

[9] S. Adhikari, N. Mithulananthan, A. Dutta, and A.J. Mathias, "Potential of Sustainable Energy Technologies under CDM in Thailand: Opportunities and Barriers", *Renew. Energy,* vol. 33, no. 9, pp. 2122-2133, 2008. [http://dx.doi.org/10.1016/j.renene.2007.12.017]

[10] E. Drury, M. Miller, C.M. Macal, D.J. Graziano, D. Heimiller, J. Ozik, and T.D. Perry IV, "The Transformation of Southern California’s Residential Photovoltaics Market through Third-Party Ownership", *Energy Policy,* vol. 42, pp. 681-690, 2012. [http://dx.doi.org/10.1016/j.enpol.2011.12.047]

[11] C.J. Koroneos, and E.A. Nanaki, "Life Cycle Environmental Impact Assessment of a Solar Water Heater", *J. Clean. Prod.,* vol. 37, pp. 154-161, 2012. [http://dx.doi.org/10.1016/j.jclepro.2012.07.001]

[12] S. Stevanović, and M. Pucar, "Financial Measures Serbia Should Offer for Solar Water Heating Systems", *Energy Build.,* vol. 54, pp. 519-526, 2012. [http://dx.doi.org/10.1016/j.enbuild.2012.02.054]

[13] K-C. Chang, W-M. Lin, T-S. Lee, and K-M. Chung, "Subsidy Programs on Diffusion of Solar Water Heaters: Taiwan’s Experience", *Energy Policy,* vol. 39, no. 2, pp. 563-567, 2011. [http://dx.doi.org/10.1016/j.enpol.2010.10.021]

[14] A. Gastli, and Y. Charabi, "Solar Water Heating Initiative in Oman Energy Saving and Carbon Credits", *Renew. Sustain. Energy Rev.,* vol. 15, no. 4, pp. 1851-1856, 2011. [http://dx.doi.org/10.1016/j.rser.2010.12.015]

[15] R. Alazraki, and J. Haselip, "Assessing the Uptake of Small-Scale Photovoltaic Electricity Production in Argentina: The PERMER Project", *J. Clean. Prod.,* vol. 15, no. 2, pp. 131-142, 2007. [http://dx.doi.org/10.1016/j.jclepro.2005.12.015]

[16] N.K. Sharma, P.K. Tiwari, and Y.R. Sood, "Solar Energy in India: Strategies, Policies, Perspectives and Future Potential", *Renew. Sustain. Energy Rev.,* vol. 16, no. 1, pp. 933-941, 2012. [http://dx.doi.org/10.1016/j.rser.2011.09.014]

[17] Y. Zhang, J. Song, and S. Hamori, "Impact of Subsidy Policies on Diffusion of Photovoltaic Power Generation", *Energy Policy,* vol. 39, no. 4, pp. 1958-1964, 2011.

[http://dx.doi.org/10.1016/j.enpol.2011.01.021]

[18] D. Ibarra, and C.F. Reinhart, "Daylight Factor Simulations-How Close Do Simulation Beginners 'Really' Get?", In: *Building Simulation,* vol. 196. 2009, p. 203.

[19] Egypt Energy Dashboard, "Global Energy Network Institute (GENI)", *Available: http://www.geni.org/globalenergy/library/energy-issues/egypt/index.shtml.*[Accessed: 29-Jun-2015] [Online]

[20] Energy Data and Analysis-EGYPT, "U.S. Energy Information Administration", *Available: http://www.eia.gov/beta/international/analysis.cfm?iso=EGY,* 2015.[Accessed: 29-Jun-2015] [Online]

[21] Solar Radiation Maps, "SolarGIS", *Available: http://solargis.info/doc/free-solar-radiation-maps-DNI.*Accessed: 29-Jun-2015 [Online]

[22] DESERTEC Foundation, "DESERTEC Foundation", *Available: http://www.desertec.org/ concept/.*Accessed: 29-Jun-2015 [Online]

[23] CO2 Emissions in Egypt, "Worldbank", *Available: http://data.worldbank.org/ country/egypt-ara- -republic?display=default.*Accessed: 29-Jun-2015 [Online]

[24] Strategy and Mandate, "New and Renewable Energy Authority", *Available: http://www.nrea.gov.eg/ english%5Cpage61e.htm.*Accessed: 29-Jun-2015 [Online]

[25] K.D. Patlitzianas, "Solar Energy in Egypt: Significant Business Opportunities", *Renew. Energy,* vol. 36, no. 9, pp. 2305-2311, 2011. [http://dx.doi.org/10.1016/j.renene.2011.03.006]

[26] "Rets. I. Gouvernement du Canada, Ressources naturelles Canada, Secteur de l'énergie, Centre de la technologie de l'énergie de CANMET – Varennes and R. and B. P. Government of Canada, Natural Resources Canada, Energy Sector, Energy Technology Branch, CANMET Energy Diversification Research Laboratory", In: *RETScreen International.*

[27] NASA, "NASA Surface Meteorology and Solar Energy", *Available: https://eosweb.larc.nasa.gov/ sse/.*Accessed: 16-Aug-2015 [Online]

CHAPTER 9

Heating Poultry Houses Using Renewable Energy

Heidi El Zanaty, Mohamed El-Morsi and **Salah El-Haggar***

The American University in Cairo, Department of Mechanical Engineering, Cairo, Egypt

Abstract: In broilers poultry house, fuel-based heating systems are commonly used to maintain the targeted temperatures for successful breeding of chicken. A considerable amount of fuel is consumed for this application, which leads to high running cost and contributes to the increase of air pollutant emissions. To investigate the economic feasibility of using solar energy as a substitute for the heating of poultry houses, a case study is conducted on a poultry broiler house located in El-Menia in Egypt. The location is selected based on the controversy foreseen due to the availability of high solar radiations versus the low fuel prices and absence of tax credits in Egypt. The case study includes the technical and economic study for a solar heating system (SHS) and the integration of biogas produced from chicken manure as an auxiliary source of heat. The heating demand is calculated hourly over a complete year using TRNSYS simulation tool. Accordingly, a SHS is designed to cover part of this demand besides a fuel based auxiliary source. The two main design variables of the SHS are the area of the solar collector and the volume of the storage tanks. An economical study of the SHS is carried out, where the net present value is calculated. The calculation is performed using the Egyptian and international fuel price.

Keywords: Bio-digesters, Biogas, Case study, Chicken manure, Economic study, Egypt, Energy, Fuel saver, Heating, Infiltration, Latent heat storage, Life cycle savings, Net present value, PCM, Poultry, Renewable, Simulation, Solar, Solar collectors, Storage tanks, Target temperature, TRNSYS.

INTRODUCTION

Poultry cindustry is one of the energy intensive industries that consume large

* **Corresponding author Salah El-Haggar:** Department of Mechanical Engineering, School of Sciences and Engineering, The American University in Cairo, AUC Avenue P.O. Box 74, New Cairo 11835, Egypt; Email: elhaggar@aucegypt.edu.

Salah El-Haggar, Sarah Mousa & Mohamed El-Morsi (Eds.)
All rights reserved-© 2016 Bentham Science Publishers

quantities of fuel, especially for the Broilers sector. Broiler poultry houses – producing chicken meat – heavily consume diesel fuel, gasoline or gas for their heating systems in order to maintain the temperatures required for the breeding of the chicks. The required temperature of any broiler house ranges from 22°C to 32°C depending on the birds' age. At each stage of the birds' development there is one optimum temperature zone in which the birds make the best performance in terms of the use of feed energy for growth and meat yield. If the birds are kept at a temperature that is lower than the targeted optimum temperature, the birds increase their feed intake and use more of the feed energy to keep their bodies' warm. This increases the production cost and decreases the meat yield. While, if kept at a temperature that is higher than the targeted optimum temperature, they reduce their feed intake to limit heat production. This also results in lower meat yield [1]. Therefore, an efficient heating system is necessary for any broiler house to maintain the required temperatures all around the year.

Currently, the heating systems used depend either on electricity, gas, diesel fuel, kerosene or other non-renewable sources. The equipment widely used in poultry houses are either direct hot air generators or indirect hot air generators-with external exhaust. The principle of operation of a hot air generator is having a pump that drives fuel under pressure to a burner nozzle. The fuel is sprayed into the combustion chamber, where it is burnt. The produced hot exhaust gases are used to heat an air stream that is supplied by a motor fan. The running cost of the aforementioned equipment is high and affects the economic efficiency of poultry houses.

In addition to its high running cost, conventional fuel-based heating systems contribute to the increase of air pollutant emissions. CO_2 emission is one of the main sources of global warming that the world is suffering. Globally, space heating and cooling systems consume 30–50% of global energy consumption, which is 5.6×10^{10} MWh/year corresponding to the emission of 1.4×10^{10} tons CO_2 per year [2]. A study made on the greenhouse gas emissions in EU countries showed that poultry produces 1.6 kg CO_2-eq per kg of chicken [3]. Clearly, the amount of CO_2 emitted per chicken varies from case to case depending on the amount of fuel used.

The amount of fuel used for heating a poultry house depends on many factors. One of the main factors is the climatic conditions [4]. The location of the house defines the annual mean air temperature and the lowest air temperatures reached during winter time, and thus determines the amount of space heating needed inside the house. Also, the level of insulation of the poultry house building indicates the amount of heat leakage to the outer atmosphere. Another factor affecting the amount of fuel used is the amount of heat produced by the chicken within the space to be heated. Chicken produce latent and sensible heat depending on their weight, age and brooding temperature [5]. Thus, the number of birds housed per volume and their weight affects the amount of heat needed to maintain a certain set temperature at this volume.

Many fields of research have been explored for utilizing different states of renewable energy including, geothermal, wind, solar and biomass energy. Solar energy is considered one of the most promising sources of renewable energy. The total annual solar radiation received on earth is approximately 3,400,000 EJ, which is tremendously greater than all other discovered and undiscovered non-renewable energy resources worldwide and is thousands of multiples of the world's total annual primary energy consumption of 450 EJ [6]. Being one of the most abundant sources of renewable energy, there is a great opportunity to utilize solar energy for space heating of poultry houses. Solar energy is extensively investigated worldwide to be used in different thermal applications such as solar water heaters, driers (air heaters), cookers, ponds, architecture, air-conditioning, chimneys and power plants [6]. Space heating of poultry houses can be achieved using the developed solar air or water heating techniques. However, it has to be noted that one of the main challenges of using solar energy is the energy storage method. Generally, for thermal energy storage, energy is stored by the change in the internal energy of a material by sensible heat, latent heat or thermo-chemical heat [7]. Sensible heat storage depends on raising the temperature of a certain material, whether solid or liquid, to utilize the heat capacity of this material. The amount of energy stored, Q_S, is determined by the specific heat of the material, C_p, the mass of the storage material, m, and change in temperature as shown in Equation (9-1)

$$Q_S = \int_{T}^{T_f} mC_p dT \tag{9-1}$$

where, T_i and T_f are the initial and final temperatures [8]. As for the latent heat storage (LHS), it depends on the heat absorption or release of a material when it changes from solid to liquid or liquid to gas or vice versa. That material used for latent energy storage is called phase change material (PCM). When the temperature rises, the chemical bonds within the PCM break up as the change of phase occurs; this can be described as an endothermic process absorbing heat. When the phase change temperature of the PCM is reached, the material starts to melt and the temperature is kept constant until the melting process is completed. The heat stored during the melting process is called the latent heat [7]. The amount of energy stored, Q_L, is shown by Equation (9-2).

$$Q_L = ma_m \Delta h_m + \int_{T_i}^{T_m} mC_p dT + \int_{T_m}^{T_f} mC_p dT \tag{9-2}$$

where a_m is the fraction melted; Δh_m is the heat of fusion per unit mass and T_m is the melting temperature [8].

A secondary source of renewable energy that can be explored for the application of heating poultry houses is biogas. Poultry houses produce a considerable amount of manure as a byproduct of the production cycle. Each bird, in addition to meat or eggs, produces a waste mainly composed of manure, litter and feathers, with a weight ranging from 4.5 to 45 kg per year depending on the sector of poultry whether it is Broilers sector producing meat, or Layers sector producing eggs or others [9]. In many countries, chicken manure is classified as waste material and generally it is a burden on the farm owners to dispose. However, when observed from a different point of view, chicken manure is a source of carbon, nitrogen and hydrogen that can be used to produce the heating gas, methane. To produce methane, the chicken manure is put into a bio-digester, which uses bacteria without oxygen (anaerobic bacteria) to degrade the organic matter and capture methane released by the bacteria in a process called anaerobic fermentation [10]. The remaining solid waste can further be used as fertilizers for agricultural applications. The produced methane can be used back again in the poultry houses as a heating fuel.

To investigate the economic feasibility of using solar energy for the heating of poultry houses, a case study is conducted on an exemplary poultry broiler house located in El-Menia governorate in Egypt. The location is selected based on the controversy foreseen in this case. Egypt is located in one of the best areas where very high long term average of annual sum of solar irradiation is available. On the other hand, Egypt is one of the highest subsidizing countries, where the retail price of diesel is below the price for crude oil on the world market. Also, there is no tax credits for renewable energy installations. Therefore, given these contradicting factors of high solar radiations, no tax credits and low fuel prices, the feasibility of using solar energy for heating poultry houses is found to be questionable and needs to be studied.

In this chapter, an exemplary poultry broiler house is chosen as a model to analyze the economics of heating poultry houses in Egypt using solar energy. A TRNSYS [11] simulation is used to calculate the head demand of the modeled house. Then, a solar heating system design is developed using a solar collector, thermal storage system and a heat distribution system to fulfil the calculated heat demand. An economic study is made to select the best design variables of the system based on the highest calculated net present value (NPV). Further economic studies are made to evaluate the performance of the solar system by varying different parameters.

HEAT DEMAND FOR A BROILER HOUSE

Production Cycle Requirements

Before simulating the broiler house and calculating the heat demand, the production cycle of broilers and its requirements has to be well understood. In a broiler house, the production cycle starts with the housing of one day old chicks. The cycle lasts for 36-42 days when the chickens' weight reach the targeted weight for selling or slaughtering. After the 42 days, the house is evacuated from all birds to be cleaned out and then disinfected. Also, the equipment is being inspected in case any maintenance is needed. Then, the house is prepared to receive the new flock of birds and start the second production cycle. The number of production cycles per year depends on the length of the cleaning and maintenance period. Usually, it ranges between 6-7 cycles per year.

The flock performance is measured by the growth rate, feed conversion ratio (FCR), livability and meat yield [12]. These are considered the main key performance indices (KPIs) for the broiler industry. To achieve the best efficiency of converting feed into meat, birds have to live in consistent optimum environmental conditions, with temperature being the most critical factor [1]. Thus, to achieve the required KPIs these environmental conditions have to be maintained within the broiler house. These conditions change daily depending on the age and weight of the birds and may slightly vary from one breed of chicken to another. In this study, the management guide for the "Arbor Acres" breed [12] is used as a reference for the chicken requirements at every age.

Methodology

A one year simulation is carried out, using TRNSYS software [11], to model the hourly performance of the poultry house throughout the year. TRNSYS Multi-zone Building Modeling (Type56) is utilized to model the thermal behavior of the poultry house. The "energy rate" method is used to calculate the temperature inside the poultry house, T_{House}, every hour. The hourly weather data represented by the ambient temperature, $T_{ambient}$, of the selected location is fed into the simulation using a TRNSYS component. Heat gains and inputs to the broiler house is calculated and transmitted to the TRNSYS building component either using an excel sheet with Visual Basic Application (VBA) add-in or directly through the building component interface. The heat gains calculated using the excel sheet are the convective and radiant heat produced by the birds. The other inputs calculated hourly, using the excel sheet, are the set temperature, T_{Set}, required to be maintained inside the house and the ventilation rate applied for the chicken. The heat gains, the ventilation rate and T_{Set} are determined based on the age of the birds, $T_{ambient}$ and T_{House}. Other heat gains introduced to the building are the heat generated from the lighting and the movement of the workers inside the house. The geometry of the building is defined using a Google SketchUp [13] drawing. The specifications of the building in terms of the materials used and infiltration rate are defined within the building component interface. A flow chart explaining the simulation structure is presented in Fig. (**9.1**). To determine the heat demand, Q_{Heat}, the calculated T_{House} is compared to the T_{Set} every hour. If T_{House} is lower than the required T_{Set}, then Q_{Heat} is calculated and supplied to reach the targeted temperature. If T_{House} is higher than the required T_{Set}, then cooling is required.

The current technology used for cooling during the summer season is evaporative cooling. Water is sprayed on pads attached to the air extractors (ventilation fans). Running these fans in the evaporative cooling mode in summer provide the required cooling effect with minimal capital and running costs compared to a chilled water system. It is worth mentioning at this stage that cooling is out of the scope of this work.

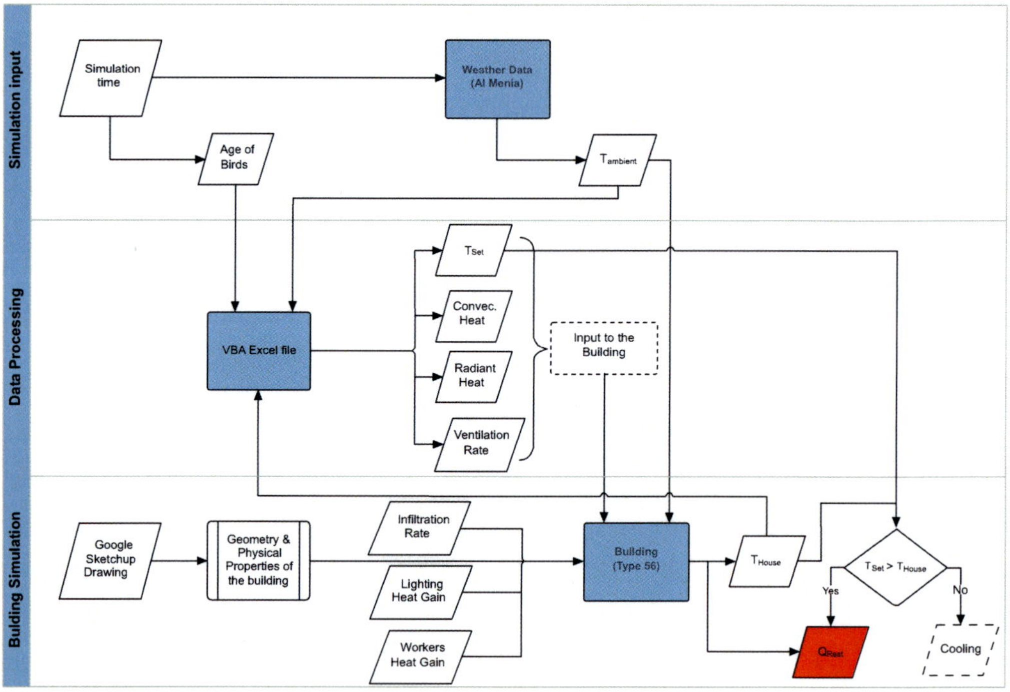

Fig. (9.1). Simulation flow chart for heat demand calculation.

Modeling the House

To set up a multi-zone building project in TRNSYS, the three dimensional building geometry is imported from the Google Sketchup drawing using the TRNSYS3d plug-in for Google SketchUp program [13]. The TRNSYS3d file is imported to the building component in TRNSYS program (TYPE56). The orientation of the building is manually defined in TRNSYS. Then, the materials of all geometrically described surfaces are defined.

Geometric Model

The Google SketchUp drawing includes the dimensions of the house, its location on Google map and the geometric definition of the different thermal zones within the house. Also, the location and dimensions of the windows (openings) of the house are defined in this drawing. An image of the model exported from Google SketchUp is shown in Fig. (**9.2**).

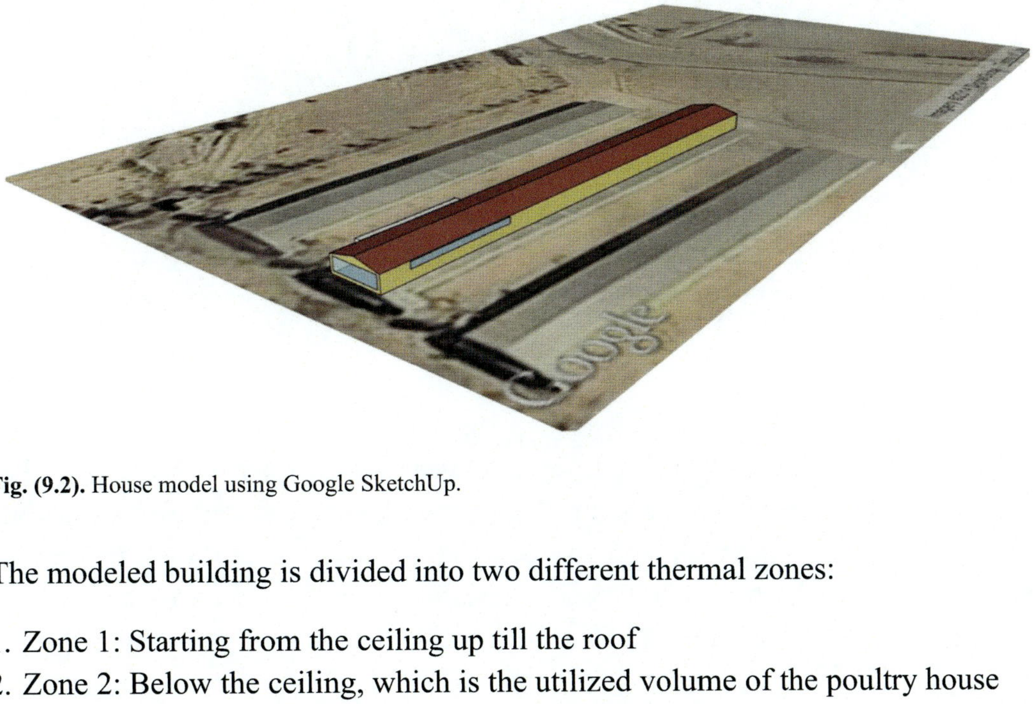

Fig. (9.2). House model using Google SketchUp.

The modeled building is divided into two different thermal zones:

1. Zone 1: Starting from the ceiling up till the roof
2. Zone 2: Below the ceiling, which is the utilized volume of the poultry house

The ceiling is the common surface connecting the two zones and it contains a thermally insulating material. This division distinguishes between the average temperature of the air above the insulation layer and that below it. As a result, this facilitates the calculation of a more accurate average temperature for the air affecting the chicken. The terminology used to describe the geometry of the house is labeled in Fig. (**9.3**). In the current example, the total area utilized by the chicken is 1,512 m^2 and the volume is 5,564 m^3.

The front end wall has an opening of 13 m wide by 2 m height. The right and left side walls have an opening of 24 m length and 1 m height. These openings are used to install a cooling system that will not be modeled as it is out of the scope of this work. The openings are closed from inside of the house using an air inlet controlled device that will be referred to as "window" hereinafter.

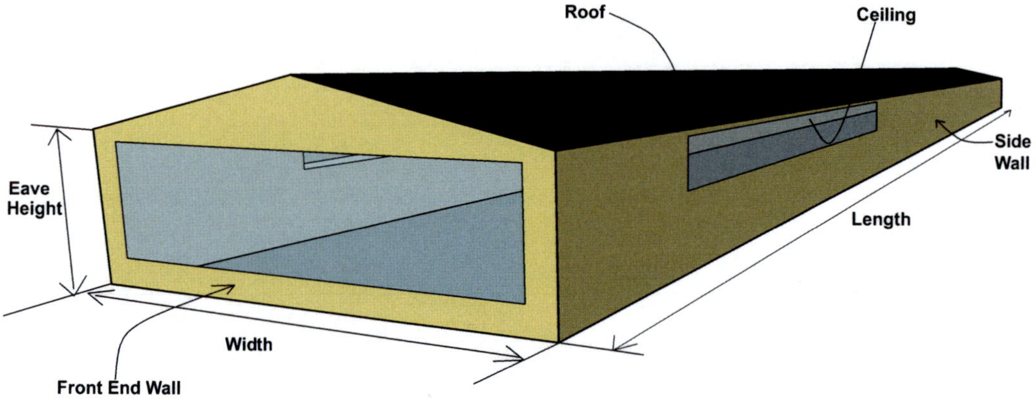

Fig. (9.3). Geometry of the modeled house.

Building Specifications

The walls are made of limestone bricks of thickness 20 cm covered by a layer of lime mortar from the outside and a layer of cement mortar from the inside. All the inside walls are painted with a layer of gypsum plastic. The ceiling consists of corrugated pre-painted galvanized sheets covered with a 10 cm thick fiberglass insulation. An air gap of about 40 cm isolates between the fiberglass and a roof made of corrugated pre-painted galvanized sheets. For windows, a window of a U-value 0.59 W/m^2K is selected from the TRNSYS library. This is the nearest value to the actual windows installed, which are made from 40 mm thick polyurethane panels of a U-value 0.5 W/m^2K [14].

Infiltration

According to the ASHRAE Handbook of Fundamentals [15], "infiltration is determined by the extent and distribution of leaks over the building envelope and the pressure differences across these leaks." To determine its value, the air exchange rate must be measured over a range of weather and equipment operation. Being a well-insulated building and a closed system poultry house, it is expected to have low infiltration rates for the modeled house. However, it is found that the fans (air extractors) used in the poultry industry are a major source of air leakage. The modeled house is equipped with 13 fans each of dimension 140×140 cm. The fans have shutters that open when the fan is running and close by gravity when it is idle.

Recent tests have shown that even in high quality and new fans, these shutters do not close tightly, which allows for air leakage that costs several hundreds of dollars in heat losses per house [1]. To evaluate the amount of air leakage, agricultural engineers from Auburn University in Alabama tested different types of high-quality fans in a laboratory under different negative static pressures [16]. The results of their test are summarized in Table **9.1**. The results of the tested fans show that the amount of air leakage per fan is significant. For the modeled house, the air leakage amount is estimated to be equivalent to that of the Aluminum shutters at 0.1 inch static pressure. The total air leakage is calculated to be 6714 m^3/hr which is equivalent to 1.19 air changes per hour.

Table 9.1. Fan shutter air infiltration test results [16].

48-inch Shutter Type	Static Pressure (inches of water)	Measured Air Leakage per Shutter (cfm)
Aluminum	0.08	270
	0.10	304
PVC	0.08	301
	0.10	351

Birds Requirements

As mentioned beforehand in section 0, there are certain indoor environmental conditions that have to be maintained in a broiler house. These conditions are changing daily throughout the production cycle depending on the requirements of the chicken. These requirements may slightly vary from one breed to another. In order to run the simulation, the targeted environmental conditions at every hour for the "Arbor Acres" breed are defined.

Set Temperature

For the Arbor Acres breed, the recommended optimum target temperature at every age is tabulated in their management guide [12] and is shown Table **9.2**. The target temperatures are given for the age from 1 day old till 27 days old with an interval of 3 days. However, the target temperature should be set to decrease gradually during these 3 days. The required input to the TRNSYS simulation is the hourly set temperature of the house. Therefore, the targeted temperature is interpolated from the values recommended by the broilers management guide to have the daily set temperature. From 27 days old till the end of the cycle, 42 days old, the temperature is maintained at 20°C [12].

Table 9.2. Arbor Acres recommended target temperatures [12].

Age (Days)	Target Temperature (°C)
1	30
3	28
6	27
9	26
12	25
15	24
18	23
21	22
24	21
27	20

Minimum Ventilation Rate

The minimum ventilation rate is "the quantity of air required per hour to supply sufficient oxygen to the birds and maintain air quality" [12]. The chicken produces carbon dioxide and ammonia by respiration and manure excretion, respectively. Therefore, the minimum ventilation is required to extract the unwanted pollutants from the house or reduce their concentration and introduce fresh air for the birds at the early age and during cold weather. At older age or during hot weather, more ventilation is used to maintain lower temperatures inside the house. Consequently, lower pollutant concentrations will be maintained. The minimum ventilation rate needed in a broiler house to maintain acceptable concentrations of ammonia and inhalable dust is 360-400 m^3/hr per 500 kg live weight broilers [17]. A more specific and weight dependent minimum ventilation rates are recommended by the Arbor Acres management guide [12]. To calculate the corresponding minimum ventilation rate per bird at every age, the daily weight of the birds is determined from Arbor Acres Broiler Performance Objectives manual [18]. Also, the total number of birds in the house and the mortality rate is considered throughout the production cycle. Generally, one of the lowest rates of mortality for broilers can be 2.5% [19]. However, this figure may vary from one farm to another. In the current study, a mortality rate of 3.7% is considered [20]. So, at the beginning of the cycle, the number of birds start with 24000 birds and the cycle ends with 23112 birds.

Heat Gains

In order to calculate the heat demand of a broiler house, it is crucial to define the internal heat gains within the volume of the house. The main effective heat gain is the one produced by the birds inside the house. Other heat gains are due to the internal lighting and workers entering the house. The heat produced from electric appliances inside the house is neglected as the only appliances found are small motors for the feeding system. All the heat gains are calculated inside "Zone 2", which is the house zone. The roof and ceiling zone, "Zone 1", has no internal heat gains. The internal heat gains are defined in the building component using the gains input window in TRNSYS, as shown in Fig. **(9.4)**. The heat gain produced by the birds is calculated hourly using the VBA excel sheet and fed into the building component as an input called "BIRDSGAIN".

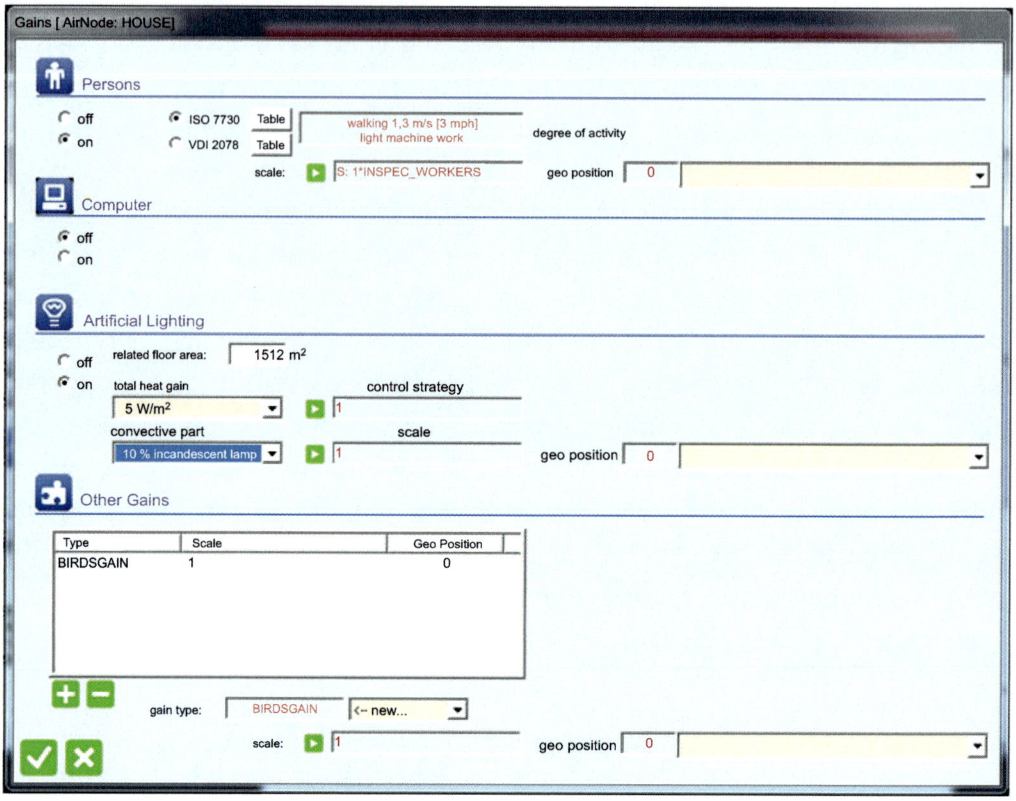

Fig. (9.4). Heat gains input window in TRNSYS.

Birds Heat Production

According to the Broilers management guide [12], the normal body temperature of a broiler chicken is 41°C and it regulates its body temperature by two methods; sensible and insensible heat loss [12]. When the ambient temperature is between 13–25°C, sensible heat loss occurs as physical radiation and convection to the cooler environment. When the ambient temperature rises above 30°C, insensible heat loss occurs through evaporative cooling and panting and increased respiration rate [12]. Also, part of the heat loss that is considered from broilers chicken is released by the litter. Therefore, the heat produced from broilers can be classified as heat and moisture, or sensible heat and latent heat. Gates *et al.* [5] presented empirically derived mathematical relationships which described broiler heat production as a function of age with adjustments for partitioning of sensible and latent heat production (SHP and LHP) at different brooding temperatures. Considering the different set temperature at every age (x) and the number of birds, the value of sensible heat produced (SHP) and latent heat produced (LHP) by the birds at each day is calculated using the set of equations from Equation (9-3) to (9-7).

$$\text{SHP} = K exp(-6.5194 + 2.9186x - 0.24162x^2) \quad 3 \leq x \leq 5 \quad \textbf{(9-3)}$$

$$\text{SHP} = K exp(1.8662 + 0.054213x - 0.00161x^2) \quad 6 \leq x \leq 19 \quad \textbf{(9-4)}$$

$$\text{LHP} = K\,(-42.961 + 27.415x - 2.84344x^2) \quad 2 \leq x \leq 5 \quad \textbf{(9-5)}$$

$$\text{LHP} = K\,(36.424 - 2.8998x + 0.08676x^2) \quad 6 \leq x \leq 15 \quad \textbf{(9-6)}$$

$$\text{LHP} = K(15.812 - 0.22611\,x) \quad 16 \leq x \leq 19 \quad \textbf{(9-7)}$$

For temperature t = 21.1°C [5], Equations (9-8) and (9-9) are used.

$$\text{SHP} = K\,(36.070 - 2.3107x + 0.058862x^2 - 0.00051x^3) \quad 20 \leq x \leq 39 \quad \textbf{(9-8)}$$

$$\text{LHP} = K(11.221 + 0.40495x - 0.02727x^2 - 0.000353x^3) \quad 20 \leq x \leq 43 \quad \textbf{(9-9)}$$

In Equations (9-3) to (9-9) K=0.64631 for W/kg.

Lighting

The modeled broiler poultry house is equipped with four longitudinal lines of lighting. Each line is of 108 m length and has a 60 W incandescent lamp placed every 4 meters. Therefore, the total number of lamps is 108 lamps, which is equivalent to 6480 W. The related floor area is equivalent to 1512 m^2. Thus, it is approximated to have an artificial lighting of total heat gain of 5 W/m^2. According to the broilers management guide, the lighting program for the first 7 days of the cycle is 23 hours of light and one hour of dark. However, the EU Broiler Welfare Directive [21] requires a 6 hours darkness, with at least one continuous period of darkness for a minimum of 4 hours. However, in practice, many farm managers apply a 24 hour light program to stimulate the birds to eat more. Following the common practice, in this study, the artificial lighting is set to be on continuously.

Workers Heat Production

In the modeled broiler poultry house, daily inspection is done by workers throughout the broiler cycle. The inspection is done by walking around the house to collect any dead birds and check on the feeding and drinking systems. Using schedule type manager, the inspection schedule is defined such that starting 6 a.m. every 6 hours one worker will enter the house to do the routine inspection work. The rate of heat gain from the occupants inside the house depends on the level of activity done. It is assumed that the inspection work in a poultry house is equivalent to walking at a speed of 1.3 m/s or doing light machine work activity as defined by the ASHRAE Standards for heat gain calculation concepts. At this level of activity, the sensible heat is equal to 100 W and the latent heat is equal to 205 W.

Heating Demand and TRNSYS Simulation

After collecting and calculating all the required data and setting up the building using TRNSYS Multizone Building Modeling, the different components are compiled and connected together in a TRNSYS project to run the simulation. Fig. (**9.5**), shows all TRNSYS components used for the simulation. The broiler poultry house "Building" is modeled using TRNSYS (Type 56) component. (TYPE 56) is an energy balance model that contains the building geometric model, building specifications and infiltration rate. In this model the convective heat flux is calculated by summing up the convective gain from the surfaces, the infiltration gains, the ventilation gains and the internal convective gains by people and lighting [22].

The TRNSYS weather data generator is used to generate the hourly weather data for the selected location. The weather data is fed to the building model and the VBA excel file. The excel file is used to calculate the required set temperature, minimum ventilation rate and birds SHP and LHP. The data calculated in excel is fed to the building where the transfer function relationships of Mitalas and Arseneault [22] are used to model the walls of the building and calculate the heat conduction through the walls. The long-wave radiation exchange between the surfaces within a zone and the convective heat flux from the inside surfaces to the air of that zone are approximated using the star network approach given by Seem [22].

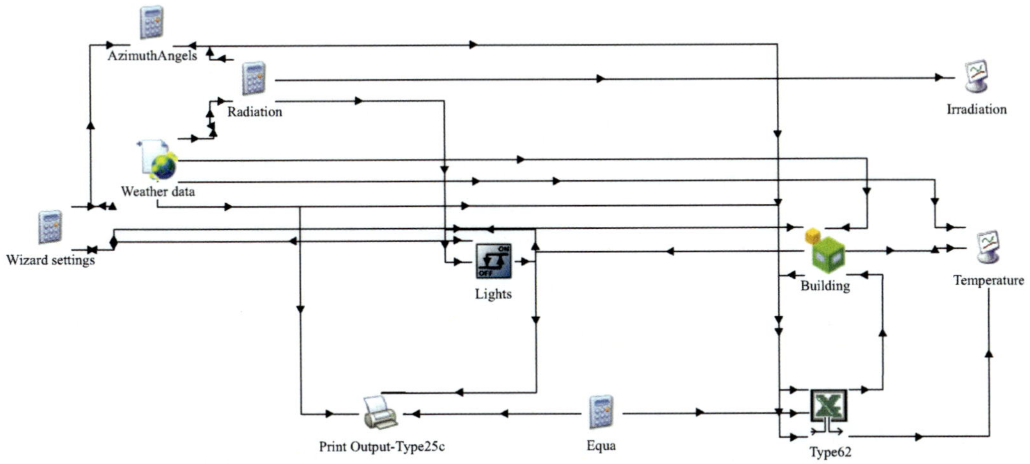

Fig. (9.5). TRNSYS Simulation for heat demand calculation.

Simplified heating is used to determine the energy requirement in an idealized way, *i.e.* the heating of the house is set to be automatic with unlimited energy source to fulfil the target temperature. Heating equipment is simulated to supply heating power that is a function of the zone temperature. If T_{House} at "Zone 2" is within the heating region at the end of a time-step, heating is applied throughout the time-step so that the final T_{House} just reaches T_{set}. By this technique, Q_{Heat} is calculated in kJ/hr at every time-step of the simulation. Similarly, simplified cooling is imposed to avoid T_{House} higher than T_{set}. This will not be modeled as it is out of the scope of this work.

Fig. (**9.6**) shows the heating demand, Q_{Heat}, required to maintain the house at a temperature $T_{House} = T_{set} - 0.5°C$, where –0.5°C represents the "dead band" of the controller. The maximum Q_{Heat} is observed on the 1st day of the first cycle, 1st of January. Also, Fig. (**9.6**) shows that the least heating demand is during the fourth and the fifth cycles, which are the summer cycles running from June to August. Only at the beginning of the cycle heating is always required to maintain the 33°C set temperature.

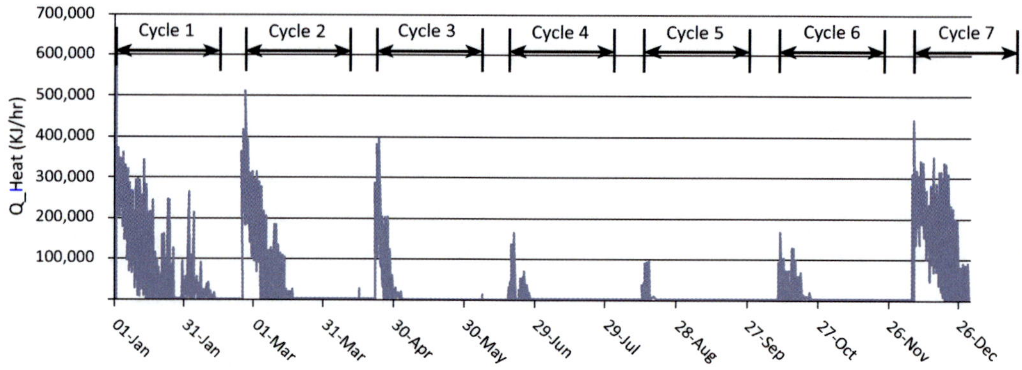

Fig. (9.6). One year heat demand simulation Q_{Heat}.

DESIGNING A SOLAR HEATING SYSTEM FOR POULTRY HOUSES

The Structure of the Solar Heating System

A solar energy system is used for heating the broiler house. The solar heating system (SHS) consists of the following three linked subsystems: (a) solar thermal collector, (b) thermal storage system and (c) heat distribution system, as illustrated in Fig. (**9.7**). The solar thermal collector is where the solar radiation is received and transmitted into the system *via* the heat transfer fluid. The thermal storage sub-system is where the solar energy produced from the collectors in the form of thermal energy is stored to be available for the varying all-day demand of the house. The heat distribution system is where the thermal energy is extracted from the heat transfer fluid and delivered to the broiler house.

The three subsystems are linked together to form two fluid closed loops. The first is the solar loop where the heat transfer fluid circulates between the solar collector and the thermal storage system. The second is the heat distribution loop where the heat transfer fluid circulates between the thermal storage system and the heat distribution system. The details of each subsystem are explained in the following section.

The design of the SHS starts by the selection of a suitable heat distribution system to provide the final product, which is thermal energy, to the house. According to the requirements of the selected system, the remaining sub-systems, which are the solar thermal collector and the thermal storage system, are designed. The design variables of the system are defined to be the area of the solar collector (A_{SC}) and the volume of the storage tanks (V_{tank}). TRNSYS simulations are run for each set of design variables to calculate the output of the SHS. Then, an economical study is performed to evaluate the performance of the system for different design variables.

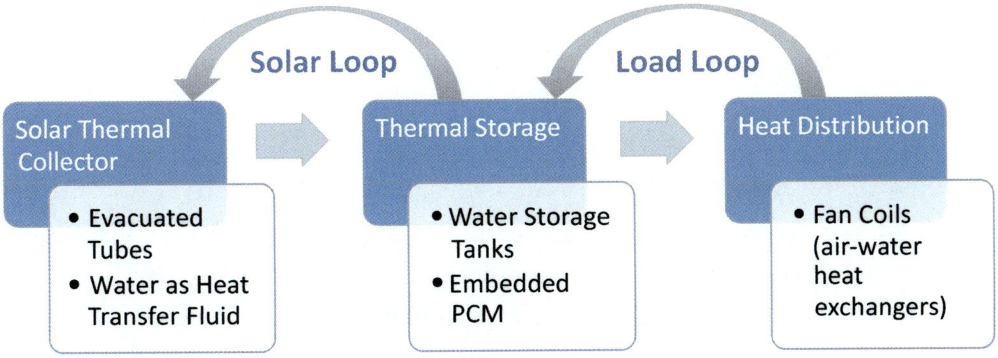

Fig. (9.7). Solar heating system (SHS) three linked subsystems.

Heat Distribution System

There are two general heat transfer methods that can be used to deliver the heat from the fluid of the SHS to the house, which are convection and radiation. Heat transfer by radiation will require installing a large amount of radiators distributed over the area of the house to maintain homogeneous temperatures within the whole volume of the house. Therefore, it is preferred to transfer the thermal energy by convection using fan coils. A fan coil unit consists of a heat exchanger that transfers the heat from hot fluid to air. Using the fan, it can generate hot air with considerable air throw to cover a larger area. Also, additional equipment like ducts and air jet diffusers can be installed on the fan coils to achieve better hot air distribution. The capacity of the fan coil units is selected based on the maximum heat required.

Solar Thermal Collector

There are several types of solar collectors such as flat plate, evacuated tubes, parabolic troughs or other types that can be used in a SHS. Also, there are different working fluids that can be used as heat transfer fluids such as water, oil or air. According to the application in hand, and after selecting the fan coils as the method for heat distribution to the house, it is compulsory to use water as the heat transfer fluid. This is because, first, the selected fan coil is designed to work with water as the heat transfer fluid. Second, water is more accessible than oil and requires no special pumps or piping accessories. To match the rated capacity of the fan coil, it is required to supply the fan coils with water temperature of about 90(C. Based on this information, the evacuated tubes solar collector is selected.

Thermal Storage System

The purpose of this system is to store the thermal energy and deliver it to the heat transfer fluid as long as possible to cover the heating demand of the house. In this design, the energy storage technique used is sensible heat storage. This is done by feeding the hot water produced from the evacuated tube solar collector into an insulated water storage tank. The water is circulated from the tank to the collector in a closed loop, named as the solar loop. Similarly, the hot water is fed from the tank into the fan coils and circulated back to the tank in a closed loop, named as load loop. The volume of the tank is considered a design variable that is to be determined based on the economic study of the system. One of the factors affecting the results is the total surface area of the storage tanks, because it contributes to the heat losses to the ambient. So, for the design in this study, it is decided to install the storage tanks inside the poultry house. In this case, the heat losses from the tanks surface area are recoverable to the heated space. Also, it is decided to use a single tank configuration for the simulation regardless of the volume of the tank. However, for practical application of the design, a multi-tanks system can be used as a replacement for the single tank to accommodate for the space constraints.

TRNSYS Simulation of the Designed SHS

To observe the outcome of the SHS with any selected design variables, it is required to run a TRNSYS simulation. The TRNSYS project is set to plug in the values of the design variables and run a one year simulation for the system, and observe the results. The TRNSYS project, depicted in Section 0, is used as the starting point for building the SHS simulation. The theoretical automatic heating is de-activated and replaced by the heat supplied by the fan coil units. The outputs that are selected to be read hourly from the simulation are as follows:

1. Air temperature of the house (T_{House}) in °C.
2. Average temperature of the water in the storage tank (T_{Tank}) in °C.
3. Outlet temperature of the collector (Tout SC) in °C.
4. The sum of total heating rate for the 5 fan coil units (QFC) in kJ/hr.

The accumulated auxiliary energy needed over the year is calculated by summing up Q_{aux}. Then, it is expressed as percentage of the total heat demand of the house by dividing it by the accumulated Q_{Heat}. So, the outcome of the simulation is the percentage of auxiliary energy required relative to the overall heat demand of the house.

CONVENTIONAL FUEL-BASED HEATING SYSTEM

Fuel Consumption in Poultry Houses

To calculate the running cost of the conventional diesel fired heaters for the modeled house, it is necessary to know the consumption of fuel over the year. According to field collected data, the modeled house consumes 600 liters of diesel fuel per day during the winter season to maintain the required set temperatures [20]. Using the results obtained from the heating demand TRNSYS simulation, the accumulated energy needed for each production cycle is calculated in terms of percentage and the data obtained for the consumption of the 1st cycle (winter cycle) is used as a reference for the remaining cycles as shown in Table . From this data, it is concluded that the yearly consumption of the house is about 61,000 liters of diesel fuel per year.

Table 9.3. Estimated fuel consumption calculation.

Cycle No.	Calculated Qheat (kJ/cycle)	Normalized Percentages (%)	Fuel Consumption (Liters)
1st Cycle	87,018,020	100%	21,600
2nd Cycle	51,870,291	60%	12,875

(Table 9.3) contd.....

Cycle No.	Calculated Qheat (kJ/cycle)	Normalized Percentages (%)	Fuel Consumption (Liters)
3rd Cycle	18,708,700	21%	4,644
4th Cycle	3,045,030	3%	756
5th Cycle	1,672,980	2%	415
6th Cycle	6,403,806	7%	1,590
7th Cycle	77,518,737	89%	19,242

To confirm the data collected from the modeled farm, the analysis of cost per bird for different broiler houses are taken as reference. The analysis showed that the average consumption of diesel fuel per bird over the year, summer and winter seasons, is 0.44 liters per bird [23]. Thus, for the modeled house of capacity 24,000 birds and having 6.54 production cycles per year, the total consumption of diesel fuel for the modeled house is calculated to be about 69,000 liters per year. The average of both fuel consumption rates is calculated to be about 65,000 liters. This figure is used to determine the yearly running cost of a traditional diesel fired heaters for the modeled house in Egypt.

Fuel Prices

According to the Egyptian Cabinet Decree no. 1160 of 2014 for modifying fuel prices that is published in the Egyptian Official Gazette on 5th of July 2014, the price of diesel fuel is 1.8 Egyptian pounds per liter [24]. Using the current official exchange rate, the price is calculated to be equivalent to 0.25 USD per liter. According to the German Agency for International Cooperation (GIZ) GmbH, Egypt lies in the category of the high subsidies countries, where the retail price of diesel is below the price for crude oil on the world market [25]. To have a deeper look at the diesel fuel price in Egypt, Fig. (**9.8**) shows the prices from year 1991 till 2012 with comparison to the international prices [25]. The figure compares the retail price of diesel oil in Egypt to (i) the price in the United States of America, which may be considered as the international minimum benchmark price for a non-subsidized policy, (ii) the price in Luxembourg, which is recorded in November 2012 as the lowest in the European Union countries. Also, the figure depicts the price of crude oil on world market as a benchmark.

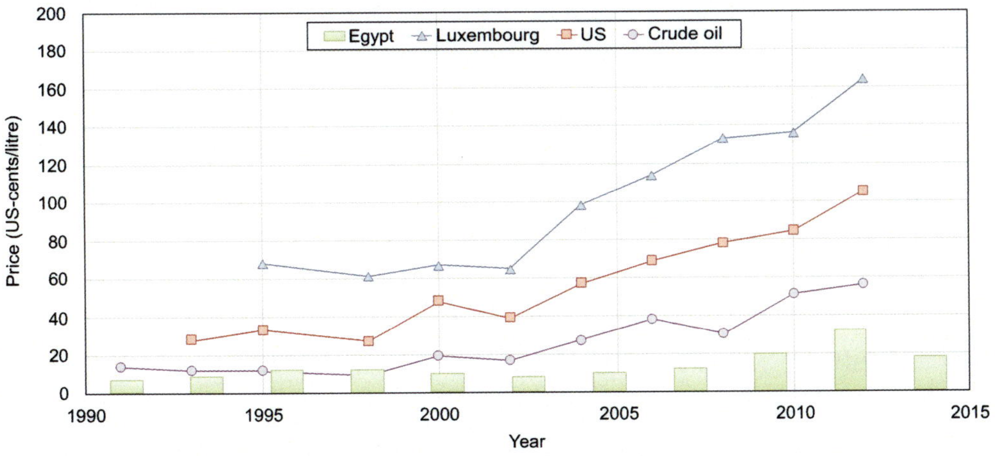

Fig. (9.8). History of retail price of diesel oil (1991-2012) [25].

In this study, the equivalent for the Egyptian local official price of diesel fuel in USD is used for the economic study of the designed SHS. Thus, the price is considered to be 0.25 USD/liter. A further comparative study is carried out using the international minimum benchmark price, which is equal to 1.05 USD/liter as of November 2012.

PRICING OF THE SOLAR HEATING SYSTEM

To be able to assess the investment value of the system, a research is done to collect the prices of the different components, from different local and international suppliers. The main constituents of the SHS are: fan coil units, water storage tanks, evacuated tubes solar collectors and pumps. Other expenses, listed below, have also to be considered when pricing the system.

Piping and Accessories

This item includes the piping network interconnecting the solar collectors together to the storage tank as a closed loop as well as the other loop connecting the storage tank to the fan coils. All the pipes connecting the solar collectors up till the tank have to be well insulated. The network includes all connections and safety equipment such as ball valves, relief valves, air vents, non-return valves …*etc.* The cost of this item is estimated to be $600 as fixed cost that does not depend on the size of the system in addition to \$35 per 1 m^2 of solar collectors.

Liquid Additive

To increase the lifetime of the system and protect the components, anti-rust additives are used. An estimated cost that depends on the storage volume is considered to be 410 \$/m^3 of storage tanks.

Electric Material

For any solar heating system, regardless of its size, cables, cable trays and a control unit will be installed to operate the system. The cost of the electric material and wiring of this system is also estimated to be \$1,000.

Installation Fees

For the installation of the system, it is estimated to have a fixed cost of \$1,000 in addition to a size dependent cost of \$15 for every m^2 of solar collectors.

In most of the solar heating systems quotations, there is a considerable cost for the construction of the supports for mounting the solar collectors. However, in this study, the solar collectors are mounted on the existing roof of the poultry house. Thus, this cost is saved as the roof is readily inclined to fulfill the requirements of the collectors' installation instructions. The total cost of the system is calculated based on Equation (9-10).

$$I.C = 4450 + 2800F.C. + 1810V_{Tank} + 425A_{SC} \quad \textbf{(9-10)}$$

where, I.C is the investment costs in United States dollars (\$), F.C is the number of fan coils, V_{Tank} is the storage volume of the tanks in (m^3) and A_{sc} is the area of the solar collectors (m^2). Table **9.4** depicts a breakdown for the items cost for of an example SHS.

Table 9.4. Summary of investment cost for a SHS.

	Item	Description	Unit	QTY	Unit Price (USD)	Total Price (USD)
1	Fan Coils	40 kW fan coils	Number	5	\$2,800	\$14,000

(Table 9.4) contd.....

	Item	Description	Unit	QTY	Unit Price (USD)	Total Price (USD)
2	Storage Tanks	2 m^3 buffer tank	m^3	6	$1,400	$8,400
3	Solar Collectors	Evacuated tubes	m^2	70	$375	$26,250
4	Pumps	Solar + Circulating	Number	2		$1,850
5	Others:					
5.1	Piping & Accessories					$3,050
5.2	Liquid Additives					$2,460
5.3	Electric material					$1,000
5.4	Installation Fees					$2,050
	Total Investment					**$59,060**

ECONOMIC STUDY METHODOLOGY

The following section illustrates the procedures followed to study the economics of the SHS. Given a set of values for the design variables, which are the area of the solar collector and the volume of the storage tanks, it is possible to

1. Run TRNSYS simulation to know the percentage of auxiliary energy needed to cover the heat demand of the house.
2. Calculate the value of the yearly fuel savings.
3. Calculate the cost of investment.

To be able to select the best economical values of the design variables, A_{SC} and V_{tank}, it is required to carry on an economical study to find the best trade-off between the cost of investment and the amount of fuel saved by the system. Among the different economic evaluation criteria, the Life Cycle Savings method (LCS) is selected for this study. LCS is defined as "the difference between the life-cycle costs of a conventional fuel-only system and the life-cycle cost of the solar plus auxiliary energy system" [26]. This is done by calculating a discounted cash flow using the present worth method. This method converts all the anticipated cash flows, expenses and savings, to a single sum equivalent at time zero [27]. This single value is the Net Present Value (NPV). Any study yielding a positive NPV means the investment under study is economically acceptable.

When comparing different investment alternatives, the greatest value of NPV is for the most recommended investment. To calculate the NPV, it is required to know the discount rate at which the present worth factor (PW) is calculated over a certain period of time using Equation (9-11) [26]:

$$PW = \frac{1}{(1+d)^N} \quad \textbf{(9-11)}$$

where, d is the market discount rate and N is the cash flow period. The market discount rate takes into account the percentage of the investment that is financed by a bank loan *versus* that covered as equity (paid by the owner or shareholders). While calculating the NPV, the inflation rate and tax deductions are taken into consideration. The following sections explain the method of calculation of the NPV.

Life-Time of the System

The cash flow period (N) represents the life-time of the system on which the analysis will be made. To determine this period, it is important to evaluate the lifetime of the main components of the SHS. For the selected APRICUS evacuated tubes solar collector, their lifetime is found to be from 15-20 years [28]. As for the fan coil units, they were found to have 20 years of service life time [29]. The remaining main component, storage tanks, is estimated to have a lifetime of 20 years as well. Based on these figures, the system lifetime of the analysis is selected to be 15 years.

Loans

As mentioned above, part or even all of the investment can be financed through a bank or an institution against a certain interest rate. The interest rate of lending from the central bank of Egypt is found to be 10.25% as of July 2014 [30]. By surveying different commercial banks, the average lending interest rate is found to be 12%. Thus, given the total investment cost, the percentage of this cost that is to be financed by a loan, the interest rate of 12% and the number of payment years N=15, the annual loan payment is calculated. The loan payment is a fixed installment that is constant over the 15 years. However, it can be analyzed into two varying payments: the interest expense and the principal re-payment. The interest expense is reduced every year as the principal payment grows.

Renewable energy projects can be financed by special funds that are concerned with environmental aspects and sustainable development. These funds may be giving grants or loans at very low interest rates. However, for this research, a conservative approach is taken by considering the commercial banks as the only source of financing.

Maintenance

The cost of maintenance of the system is divided into two types: operational costs and materials (spare parts) costs. The operational cost is the extra cost of general maintenance and cleaning of the solar system in comparison to that of the conventional fuel based system. It is estimated to be $100 per month. So, a yearly expense of $1200 is included in the cash flow starting from year one.

The materials or spare parts costs is usually considered as a fraction of the capital cost. However, it is related to the warranty and lifetime of the different components of the SHS. Thus, a price breakdown is done to divide the components into 4 categories: Collectors, Tanks, Fan Coil units and others. The others category includes all other components such as the pipes, valves, accessories, electric material and liquid additives. The cost of maintenance for the four categories is considered to be a fraction of 2% from their capital cost. This cost will inflate yearly by 8.47%, which is the core inflation rate announced by the Central Bank of Egypt as of October 2014 [30]. However, the expense of each category will be actuated after the end of its warranty period. The solar collectors price includes a warranty period of 10 years, thus its maintenance expense will start from year 11. The tanks warranty is valid for 5 years, so its expense starts at year 6. The fan coils units life time is 20 years, thus it is expected to start its maintenance expenses at year 11. For the others category, the maintenance starts at the first year of operation.

Parasitic Energy

The parasitic energy, summarized in Table **9.5**, is the power required to run the components of the SHS, which are mainly the pumps and fan coil units.

Using the normalized heat demand percentages, in Table **9.3**, and referring to the number of operating hours of the conventional fuel system, the maximum total number of operating hours per year is estimated to be 2268 hours. The parasitic consumption of the SHS is calculated based on the solar fraction of the operating hours (depending on the auxiliary energy needed) multiplied by the total parasitic power.

Table 9.5. SHS parasitic power.

Item	Rated Power (kW)
Pumps	4.47
Fan Coils	2.65
Total Parasitic Power	7.12

According to the Egyptian law no. 1257 of 2014, the electricity tariff is subjected to an annual increase for the next five years [31]. To estimate the tariff after the 5 years period, an inflation of 10% is used. The tariff of interest, which is for low voltage 380V applications, is multiplied by the parasitic consumption to calculate the yearly parasitic expenses in USD as per the current official currency exchange rate.

Fuel Savings

The Egyptian government has been announcing since 2011 that it will lift the energy subsidies on industries, which is causing a lot of controversy within the business sector [32]. However, by observing the historical time series of the fuel prices in Egypt and putting in consideration the social and political pressures against the reduction in subsidies, a conservative inflation rate (i) for diesel fuel equal to 2% is considered for this study. So, the price at the first year is set to be $0.25/liter as concluded previously. The fuel savings (F) at this year is calculated based on the amount of fuel saved. The amount of fuel saved is calculated by multiplying the total yearly fuel consumption concluded in previously, which is 65000 liters, by the percentage of energy saved by the SHS. Thus, the fuel saving for the first year (F_1) is calculated using Equation (9-12) as follows:

$$F_1 = \text{Price of Fuel} \times \text{Yearly Fuel Cosumption} \times (1\text{-}\%\text{Aux. Energy Needed}) \qquad \textbf{(9-12)}$$

Then, to account for the inflation rate starting the second year, for each year N, F_N is calculated based on Equation (9-13) as follows:

$$F_N = F_1\ (1+i)^{N-1}, \text{ where } i=2\% \qquad \textbf{(9-13)}$$

Tax Savings

By law, some of the expenses are tax deductible from the Income Tax that is paid by any entity. Therefore, for an income tax (t) and a tax deductible expense C_t, the tax deduction D_t is extracted from the life-cycle cost as show in Equation (9-14) below [29]:

$$D_t = t \times C_t \qquad \textbf{(9-14)}$$

According to the income tax law of Egypt, t=25% [33]. For this research, where the house is a producing unit and not a residential home, the expenses that are tax deductible are: the loan interest expense, the maintenance cost and the parasitic energy expense.

Discount Rate

The discount rate is used to determine the present value of future cash flows, where it is considered "a consequence of the productivity of capital" [29]. It takes into account the time value of money, and also the risk or uncertainty of future cash flows. In this research, the weighted average cost of capital (WACC) is used to determine the discount rate, and is calculated using Equation (9-15) as follows [34]:

$$WACC = \frac{E}{E+D} \times k_E + \frac{D}{E+D} \times k_D \qquad \textbf{(9-15)}$$

where, E is the amount of equity used in financing the project, k_E is the after tax equity rate of return, D is the amount of debt (loan) used in financing the project and k_D is the after tax debt interest rate. k_E takes into account the risk of the investment and the alternative of depositing the amount in a risk free depositary. The 10-year bond yield of 15.28% -as per the ministry of finance, is considered as the risk free rate [35]. Considering a risk premium of 8% and a security factor (=1, the k_E is calculated to be 19.5% after tax deduction.

Based on the central bank's loan rates [30] and by surveying the commercial banks' rates, the loan interest rate is set to be 12%. Thus, the after tax rate k_D is calculated to be 9%. In this study, it is assumed that 90% of the investment will be financed through a bank loan, while the remaining 10% will be paid by the owner or shareholders. Based on this percentage, the discount rate is calculated to be 10.05%.

Salvage Value

The economic study is made on a period of 15 years as verified previously in section 0. At the last year, N=15, a re-sale value of the SHS is estimated to be 10% of the total capital cost.

RESULTS AND DISCUSSION

Fuel Saver System

In this section, the SHS is dealt with as a fuel saver system that is complementing an already existing conventional fuel based heating system. Therefore, the economic study takes into account only the additional cost of the SHS and deducts the fuel savings achieved by the SHS.

As mentioned beforehand, the solar area collector A_{SC} and the volume of storage tanks V_{tank} are the two design variables of the SHS. To examine the magnitude and direction of the effect of each variable on the performance of the SHS, a 2^2 factorial numerical experiment is carried out. The evaluation of the performance of the system is done based on the NPV, which is the outcome of the economic study. A solution space is explored to search for the best design, which yields the maximum NPV. The effect of adding latent heat storage to the designed sensible heat water storage tanks is explored. Also, the effect of reducing the infiltration rate of the building is examined.

Effect of the Design Variables on the NPV

Based on the literature, a solar system performance is much more sensitive to collector area than to any other variable [26]. A 2^2 factorial numerical experiment is carried out to confirm this statement. The two factors to be experimented are the A_{SC} and the V_{tank}. Using the initial estimate of A_{SC}= 90 m^2, the high and low values of each variable are selected to be multiples of the single collector area of 4.097 m^2 and single storage tank of volume 2 m^3. The selected values are presented in Table **9.6**.

Table 9.6. Low and high values of the design variables.

Variable	Low Value	High Value	Unit
A_{SC}	40.97	139.29	m^2
V_{tank}	4	12	m^3

Using the high and low values of the design variables, four different SHS designs are simulated using TRNSYS for a complete year. From the results of the simulations the percentage of the needed auxiliary energy is calculated as explained beforehand in section 0. Then, the fuel saving is calculated, using Equation (9-10), where the local fuel price of $0.25/liter is used. The cost of the investment is calculated using Equation (9-12). The results of the simulations and the subsequent calculations are summarized in Table **9.7**.

Table 9.7. Preliminary results of the 22 factorial numerical experiments.

Run #	A_{SC} (m^2)	V_{tank} (m^3)	Aux. Energy %	Investment (USD)	Fuel Savings (USD)
1	40.97	4	75%	43,102	4,032
2	139.29	4	19%	84,892	13,292
3	40.97	12	58%	57,582	$ 6,846
4	139.29	12	10%	99,372	14,829

For each design, the economic study is performed to calculate the response of the numerical experiment, which is the NPV of the system. The summary of the economic study for the fourth run only, where A_{SC}=139.29 m^2 and V_{tank}= 12 m^3, are illustrated here. The total energy saved in this case is 90% of the conventional system's consumption, which is equivalent to $14,828 of fuel savings at the first year. Using the fuel inflation rate of 2%, calculated discount rate of 10.05% and the salvage value at *N*=15 of $9,937, the discounted cash flow (DCF) is developed as shown in Table **9.8**. The NPV is found to be equal $11,704, which indicates that the design is economically accepted.

Table 9.8. DCF calculating NPV for A_{SC}=139.29 m^2 and V_{tank}=12 m^3.

Year	Investment Payment	Fuel Savings	Parasitic Energy Cost	Maintenance Cost	Loan payment	Interest Expense	Tax Savings	Salvage Value at N=15	Net Cash Flow	Discount Factor	NPV of Cash Flow
0	-$9,937								-$9,937	1.00	-$9,937
1	0	$14,829	-$748	-$1,527	-$13,131	-$10,732	$3,252	0	$2,675	0.91	$2,431
2	0	$15,125	-$889	-$1,554	-$13,131	-$10,444	$3,222	0	$2,775	0.83	$2,291
3	0	$15,428	-$1,073	-$1,584	-$13,131	-$10,122	$3,195	0	$2,837	0.75	$2,129
4	0	$15,736	-$1,197	-$1,617	-$13,131	-$9,761	$3,144	0	$2,938	0.68	$2,004
5	0	$16,051	-$1,359	-$1,652	-$13,131	-$9,356	$3,092	0	$3,005	0.62	$1,862
6	0	$16,372	-$1,495	-$2,027	-$13,131	-$8,903	$3,106	0	$2,832	0.56	$1,594
7	0	$16,699	-$1,644	-$2,097	-$13,131	-$8,396	$3,034	0	$2,869	0.51	$1,468
8	0	$17,033	-$1,809	-$2,173	-$13,131	-$7,828	$2,952	0	$2,881	0.46	$1,340
9	0	$17,374	-$1,989	-$2,255	-$13,131	-$7,191	$2,859	0	$2,866	0.42	$1,211
10	0	$17,721	-$2,188	-$2,344	-$13,131	-$6,479	$2,753	0	$2,820	0.38	$1,083
11	0	$18,076	-$2,407	-$3,766	-$13,131	-$5,680	$2,963	0	$1,746	0.35	$609
12	0	$18,437	-$2,648	-$3,983	-$13,131	-$4,786	$2,854	0	$1,541	0.32	$489
13	0	$18,806	-$2,913	-$4,219	-$13,131	-$3,785	$2,729	0	$1,285	0.29	$370
14	0	$19,182	-$3,204	-$4,475	-$13,131	-$2,663	$2,585	0	$972	0.26	$254
15	0	$19,566	-$3,525	-$4,752	-$13,131	-$1,407	$2,421	$9,937	$10,531	0.24	$2,505
NET PRESENT VALUE (NPV)											**$11,704**

Table 9.9a. Response values of the 22 factorial numerical experiments.

Run #	x_1	x_2				Response *y*
	A_{SC} (m^2)	**V_{tank} (m^3)**	**Aux. Energy (%)**	**Investment (USD)**	**Fuel Savings (USD)**	**NPV (USD)**
1	40.97	4	75%	$ 43,102.25	$ 4,031.64	-$ 17,630
2	139.298	4	19%	$ 84,891.65	$ 13,291.96	$ 14,508

(Table 9.9a) contd.....

Run #	x_1 A_{SC} (m^2)	x_2 V_{tank} (m^3)	Aux. Energy (%)	Investment (USD)	Fuel Savings (USD)	Response y NPV (USD)
3	40.97	12	58%	\$ 57,582.25	\$ 6,846.15	-\$ 10,436
4	139.298	12	10%	\$ 99,371.65	\$ 14,828.50	\$ 11,704

Table 9.9b. Coded variables of the 22 factorial numerical experiments.

x_1	x_2	$x_1 x_2$	y	Response Symbol
-1	-1	1	-\$ 17,630	(1)
1	-1	-1	\$ 14,508	a
-1	1	-1	-\$ 10,436	b
1	1	1	\$ 11,704	ab

The same procedure is followed to calculate the NPV value for the 3 other experimental conditions. The results are tabulated in Table **9.9**.

where, the coded variables x_1 and x_2 are defined as shown in Equations (9-16) and (9-17), respectively.

$$x_1 = \frac{A_{SC} - 90.13}{49.16} \quad \textbf{(9-16)}$$

$$x_2 = \frac{V_{tank} - 8}{4} \quad \textbf{(9-17)}$$

The symbols (1), a, b and ab are used to represent the response of the runs from 1 to 4 respectively. The contrast, which is defined as the total effect of each variable on the response, is calculated based on the Equations (9-18) to (9-20) [36]

$$Contrast_{A_{sc}} = ab + a - b - (1) \quad \textbf{(9-18)}$$

$$Contrast_{V_{tank}} = ab + b - a - (1) \quad \textbf{(9-19)}$$

$$Contrast_{V_{tank\&A_{SC}}} = ab + (1) - a - b \quad \textbf{(9-20)}$$

Then, the sum of squares, *SS*, for each contrast is calculated by Equation (9-21) [36].

$$SS = \frac{[Contrast]^2}{4n} \quad \textbf{(9-21)}$$

where n is the number of replicates, which is equal to 1 because running the same simulation for certain design variables will always yield the same response (no random effect).

The contribution of each design variable and their interaction effect on the response is summarized in Table **9.10**. The table shows that the effect of variable x_1 (A_{SC}) is the most dominant on the response with a percentage of 96.1%.

Table 9.10. The 22 factorial numerical experiment results.

Variable	Sum of Squares	% Contribution
x_1	736525321	96.1%
x_2	4818025	0.6%
$x_1 x_2$	24990001	3.3%

Solution Space

In the search for the most economical design, a solution space is explored by setting the design variables to the values shown in Tables **9.11** and **9.12**. Starting from the design center, the area of solar collectors, A_{SC}, is chosen to vary between the high and low values used in the 2^2 Factorial numerical experiment with an increment of 8.194 m^2, which is equivalent to two solar collectors. Similarly, the volume of the storage tanks, V_{tank}, will vary between the high and low values but with an increment of 2 m^3.

Table 9.11. Selected values for A_{SC} in (m^2) for the 22 factorial numerical experiment.

41	49	57	66	74	82	90	98	107	115	123	131	139

Table 9.12. Selected values for V_{tank} in (m^3) for the 22 factorial numerical experiment.

4	6	8	10	12

The 13 values of A_{SC} and 5 values of V_{tank} will yield a total of 65 combinations; in other words 65 iterations are to be explored. The objective is to find the maximum NPV within the selected solution space. The 65 TRNSYS simulations are run to calculate the required auxiliary energy, investment cost and the fuel savings. Then, the economic study is performed for the 65 solutions to calculate the NPV using the Egyptian fuel prices.

The results of the NPV study are shown in Fig. (**9.9**). The maximum NPV, iteration number 46 on Fig. (**9.10**), is where A_{SC}=114.7 m^2 and V_{tank}=4 m^3. The investment cost of this design is $74,444. At this design, the SHS covers 77% of the heat demand of the house, leaving 23% to be covered by an auxiliary source. The fuel savings at the first year is $12,631 and the NPV for the investment at this design is $19,428.

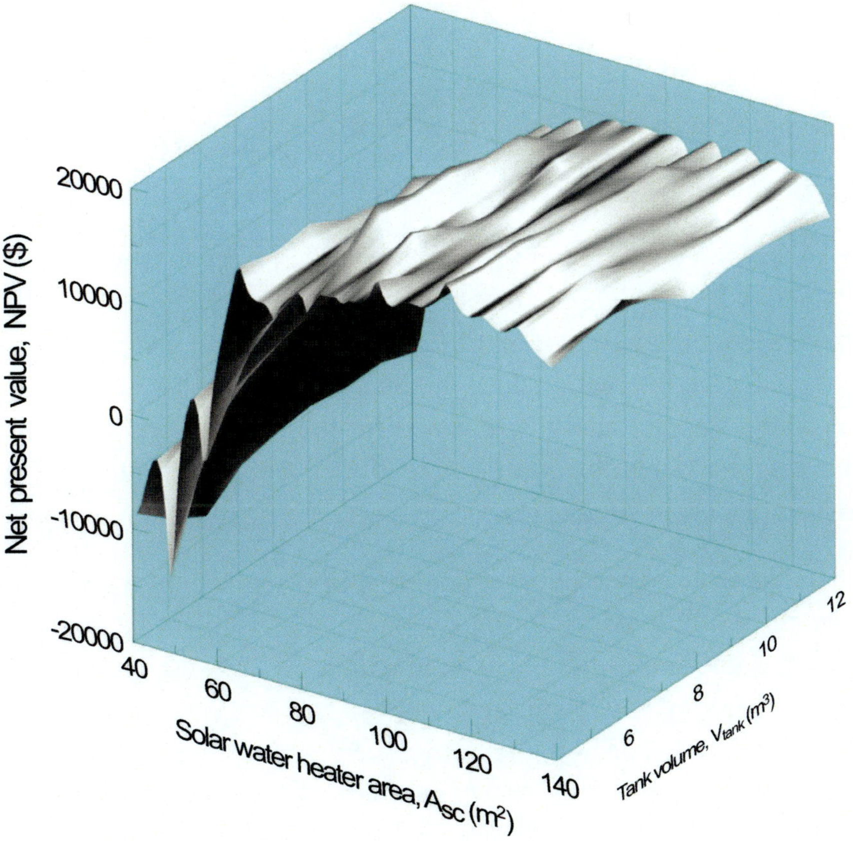

Fig. (9.9). Solution space results for NPV, using Egyptian fuel prices ($0.25/liter).

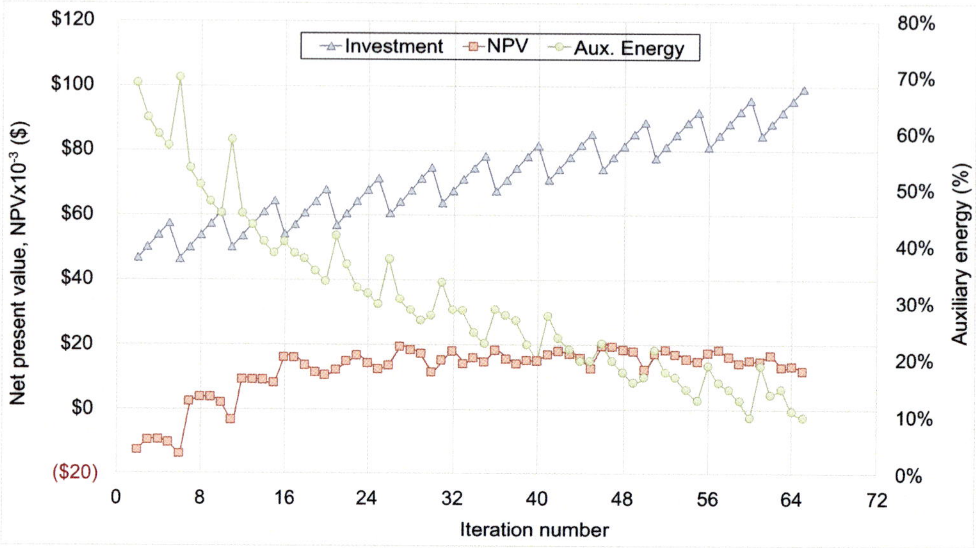

Fig. (9.10). Solution space results, using Egyptian fuel prices ($0.25/liter).

Applying International Fuel Prices

The results of the previous economic study are based on the local diesel fuel price in Egypt, which is $ 0.25/liter. As mentioned in section 0 before, the local prices in Egypt is far below the international fuel price. Therefore, a comparative study is carried out using the international minimum benchmark price, which is equal to $ 1.05/liter. The economic study is repeated for the 65 solutions and the NPV is calculated at each case based on the international fuel price. The results of the calculations are displayed (Fig. **9.11**).

The NPV calculated using the international fuel prices is found to be positive and of large value for the 65 solutions, which shows that all solutions are economically feasible and profitable. The maximum NPV, iteration number 60 on Fig. (**9.11**), is where A_{SC}=131.1 m^2 and V_{tank}=12 m^3. The investment cost of this design is $95,889. At this design, the SHS covers 90% of the heat demand of the house, leaving 10% to be covered by an auxiliary source. The fuel savings at the first year is $ 61,726 and the NPV for the investment at this design is $411,341.

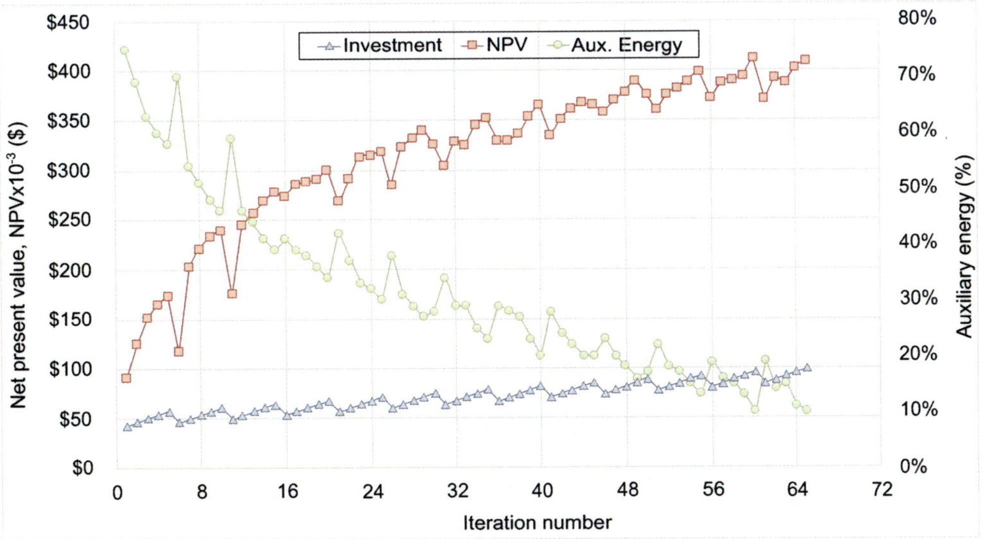

Fig. (9.11). Solution space results, using international fuel prices ($1.05/liter).

Effect of Infiltration Rate

The previous calculations are performed based on an infiltration rate of 1.19 air changes per hour (ACH), as mentioned in section 0. However, referring to the ASHRAE handbook [37], air leakage during winter time can be expressed as 0.45 ACH for a tight building with outside temperature of -1°C. The effect of infiltration on the performance of the SHS is explored. Three different designs are selected from the solution space, namely solutions number 12, 29 and 46. The performance of the three designs are evaluated by running TRNSYS simulation at 3 different levels of infiltration, namely 0.45, 1.19 and 2 ACH. This yields a total of 9 simulations. The 9 TRNSYS simulations are run to calculate the required auxiliary energy, investment cost and the fuel savings. Then, the economic study is performed for the 9 solutions to calculate the NPV using the Egyptian fuel prices. The results of this study are summarized in Table **9.13**.

Table **9.13** shows that reducing the infiltration rate yields significant savings in the required auxiliary energy for the same design variables. Thus, it yields an increase in the NPV of the investment. For the best selected design from within the solution space, solution 46, the NPV increases by a value of $7,600 (about

39% increase) when reducing the infiltration rate from 1.19 to 0.45 ACH.

Table 9.13. Results of varying infiltration rate.

Solution #	Infiltration ACH (hr-1)	Aux. Energy (%)	NPV (USD)
Solution 12	0.45	33%	$26,322.00
	1.19	46%	$ 9,128.00
	2	55%	-$ 1,900.00
Solution 29	0.45	23%	$21,647.00
	1.19	27%	$17,040.00
	2	36%	$ 4,967.00
Solution 46	0.45	17%	$27,021.00
	1.19	23%	$19,428.00
	2	33%	$ 6,492.00

Effect of Adding Latent Heat Storage Using PCM

According to the literature, latent heat storage technique should enhance the performance of the storage system due to its ability of storing energy at a constant temperature. To explore the effect of using latent heat storage on the designed SHS, the model developed by the IEA SHC TASK 32 [38], is used as a replacement for the water storage tank of the SHS. The developed (Type 840) model is a TRNSYS component that simulates water tanks with integrated PCM modules of different geometries such as cylinders, spheres and plates. One of the developers of the model, Dr. Hermann Schranzhofer [39], is contacted to acquire the model and he responded by sending all the required files to run the model using TRNSYS.

The effect of using latent heat storage on the performance of the SHS is explored only on the design of solution 46, where A_{SC}=114.7 m^2 and V_{tank}=4 m^3. A 5 m^3 water storage tank with embedded PCM solid cylinders is used as a replacement for the stratified sensible energy storage tank. The extra 1 $m^3$of volume, from 4 to 5 m^3 is used to accommodate for the addition of the PCM cylinders volume. For clarification, a sketch for the cross section of a storage tank with embedded PCM cylinders is shown in Fig. (**9.12**). In this study, D_i shown in the figure is equal to zero. The parameters of the model, TYPE 840, are set as follows:

1. Tank Volume: 5 m^3.
2. Tank Height: 2.2 m^3.
3. Number of PCM modules (cylinders): 270.
4. Diameter of PCM modules (cylinders): D_o=50 mm.
5. PCM % of cross section area from the tank: 23%.
6. Overall heat loss coefficient of the storage tank: 11.31 W/K, as shown by Equation (22) [40].

$$UA = 0.16 \times V_S^{0.5} \quad [40] \qquad \textbf{(9-22)}$$

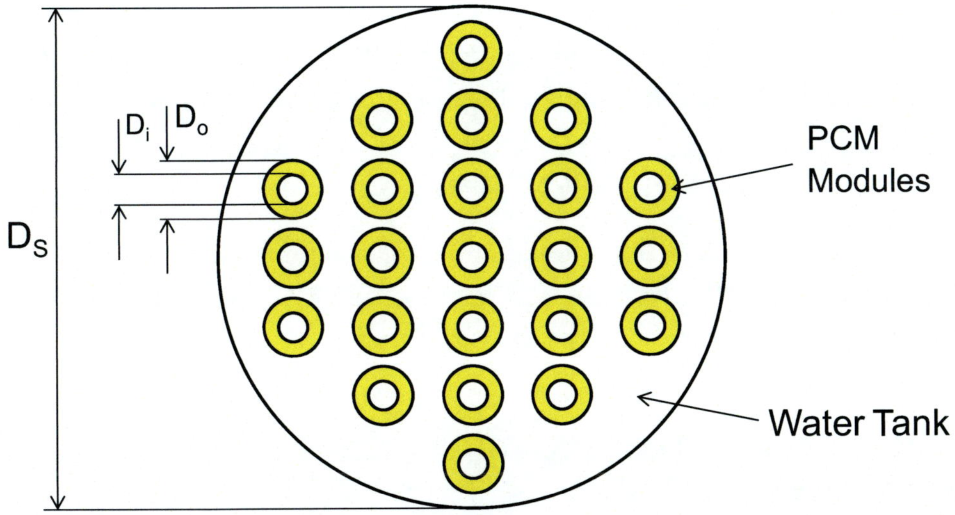

Fig. (9.12). Sketch for storage tank cross section with embedded PCM, [41].

The validation of the model done by Schranzhofer *et al.* [42] is carried out using Sodium Acetate Trihydrate with embedded graphite as the PCM material. The melting temperature of this material is between 56 and 60°C as per the material data file supplied by the author, Dr. Hermann Schranzhofer. The same material is used for the simulation of this research.

TRNSYS simulation is run using the above mentioned parameters. The results are obtained for the design of solution 46. The required auxiliary energy dropped to 21%. Thus, the replacement of the storage tanks from sensible water storage tanks to latent heat water storage tanks with embedded PCM resulted in a reduction of 2% of the auxiliary energy needed.

To evaluate the feasibility of using PCM, the NPV of the SHS after adding the PCM is calculated. First, the extra cost of adding PCM to the storage tanks is calculated. A supplier [43], is contacted to get a quotation for PCM material of similar specifications to the modelled one. The price is found to be £2.5/kg of capsulated PCM, which is equivalent to \$3.9/kg. According to the material data file, the density of the modelled PCM is 1100 kg/m^3. Thus, the weight of the PCM used is calculated to be 1,155 kg, which is equivalent to an extra cost of \$4,500. An additional cost of \$1,000 is estimated for the containers of the PCM. Therefore, the total extra cost for embedding PCM modules is \$5,500. An economic study is carried out using the local fuel price of \$ 0.25/liter to calculate the NPV. The fuel savings at the first year is \$12,945 and the total investment increased to \$79,944 and the NPV is \$15,801.

By comparing the calculated NPV to that of the SHS with sensible storage system, it is found that the NPV decreased by a value of \$3,627. In other words, for solution 46, it is more economical to use sensible water storage tanks than using latent heat storage tanks.

ALL-GREEN SOLUTION: BIO-DIGESTERS AS AN AUXILIARY SOURCE OF ENERGY

In this section, the SHS is complemented with a bio-digester to build a complete renewable all-green solution that can replace the traditional fuel based heating system of poultry houses. The anaerobic bio-digesters are used to produce biogas from litter, manure or any waste in general. The bird production of manure in a broilers house is calculated to be 1.95 kg/bird per production cycle [44]. Thus, the modeled house of capacity 24,000 broilers in this research produces 46,754 kg of manure per cycle, which is equal to about 306 tons of manure per year.

Modeled House Waste to Energy Production

According to El-Haggar [45], 7.7 kg of dry chicken manure can produce 1 m^3 of biogas and the calorific value of 1 m^3 of biogas is equivalent to that of 0.6 liter of kerosene (or diesel fuel). The educational manual of SKG Sangha [46] states that the broilers chicken manure moisture content is 25%. From this information, it can be concluded that 10.3 kg of moist chicken manure can produce 1 m^3 of biogas. Therefore, the modeled house of capacity 24,000 broilers is capable of producing 29,800 m^3 of biogas, which is equivalent to 17,877 liters of kerosene (or diesel fuel) per year.

Returning to the previous results, it is found that the required auxiliary energy for the design solution 46 is 23%. Thus, 23% of the total yearly consumption of fuel, which is equal to about 15,000 liters, is needed to complement the SHS. This being said, it is clear that the amount of fuel that can be produced from the waste of the chicken per year is sufficient to cover the auxiliary energy required to complement the SHS design of solution 46.

Based on the above, an all green solution for heating the modelled poultry house is presented by integrating the simple anaerobic bio-digesters model Deenbandhu that is introduced by SKG Sangha organization in Egypt with the designed SHS solution 46. The 23% of auxiliary energy needed for this design is equivalent to 24,950 m^3 of biogas per year. Thus, it is required to produce daily 68.3 m^3 of biogas. The largest available unit that SKG Sangha currently establishes in Egypt is a 6 m^3 bio-digester unit. Therefore, 12 units of 6 m^3 of daily production each are needed to cover the auxiliary heating demand of the modeled poultry house. It has to be noted that the produced biogas have to be stored to be used at different times of the day and the year depending on the varying heat demand.

Economic Study of the All-Green Solution

To evaluate the performance of the all green solution, an economic study is carried out to calculate the NPV of the investment. First, the extra cost of establishing the bio-digesters is calculated. According to SKG Sangha in Egypt, the cost of establishing the 6 m^3 unit is L.E10,000 [47]. Thus, the total cost for establishing the 12 units is calculated to be about $16,800. By adding this value to the cost of the SHS of solution 46, the total investment cost becomes \$91,244. The required auxiliary energy in this case will be equal to zero since all the auxiliary energy will be supplied by the bio-digesters. Therefore, 100% of the yearly fuel consumption will be saved. Using the local fuel price of Egypt of \$0.25/liter, the fuel savings at the first year becomes \$16,387. Following the same procedures, the NPV value is calculated to be \$30,237. Therefore, an additional value of \$10,800 is added to the NPV. Thus, the NPV increased by about 55% when a bio-digesters in introduced to replace diesel fuel.

Space Constraint

However, it has to be noted that the space required for the establishment of a 6 m^3 bio-digester is 4.6×4.6 m (21.16 m^2) per bio-digester unit. Therefore, a total area of 254 m^2 should be available around the poultry house to be able to establish the required bio-digester units. Also, the bio-digesters require a suitable setting for mixing the manure and waste of the poultry house with water before being fed into the bio-digesters. Finally, the handling of the discharged excess slurry and residuals of the bio-digesters should be considered.

DISCUSSION

In this work, TRNSYS is used to simulate the hourly heating demand of a poultry broiler house of capacity 24000 birds located in Al-Menia governorate in Egypt, over a complete year. The house is modeled in full details and all the heat gains are calculated and fed as an input to the TRNSYS model. The highest heating demands are found to be at the beginning of the 1st (first) and the 7th (last) production cycle.

Knowing the heating demand of the house, a solar heating system (SHS) is designed to cover part of this demand. The system consists of: Evacuated tubes as solar thermal collectors, water storage tanks as a storage system and fan coil units as heat distribution system. The two main design variables of the SHS are the area of the solar collector (A_{SC}) and the volume of the storage tanks (V_{tank}). The pricing of the system is calculated for each component and summarized using Equation (9-10). To evaluate the performance of the SHS under different design variables, the Life Cycle Savings (LCS) economic evaluation criteria is used, where the difference between the life-cycle costs of a conventional fuel-only system and the life-cycle cost of the SHS plus auxiliary energy source is calculated and presented in a discounted cash flow to determine the NPV.

The fuel consumption of the modelled house using traditional fuel-based heating system is found to be about 65000 liters per year. The local price of diesel fuel in Egypt is found to be $0.25/liter, while the international minimum benchmark price is $1.05/liter. The fuel savings achieved by the SHS is calculated based on the percentage of auxiliary energy needed using Equation (9-12). The life-time of the SHS upon which it will be economically studied is 15 years. The SHS is assumed to be 90% financed by a loan with an interest rate of 12%. The maintenance expense is defined as operational costs and materials (spare parts) costs and calculated yearly with an inflation rate of 8.5%. The parasitic energy expense is estimated yearly based on the operational hours of the system. The tax savings are calculated based on the Egyptian tax law and the discount rate is concluded to be 10.05%. A re-sale value of the SHS is set to be 10% of its cost after 15 years.

A 2^2 Factorial experiment is carried out to examine the magnitude and direction of the effect of each variable on the performance of the SHS. The high and low values of the two design variables were set to be 139 m^2 and 40 m^2 and 12 m^3 and 4 m^3, respectively. The NPV is calculated for the 4 experimental conditions. It is found that increasing the A_{SC} had 96% of the contribution to the increase in the NPV.

To determine the most economical design, a solution space consisting of 65 solutions is created by varying the design variable A_{SC} from 40 to 139 m^2 and V_{tank} from 4 to 12 m^3. The NPV is calculated at each solution using the local fuel price in Egypt, which is $0.25/liter. The best solution with highest NPV of $19,428 is found to be at solution 46, where A_{SC}=114.7m^2 and V_{tank}=4 m^3. The investment cost of this design is $74,444 and it covers 77% of the heat demand of the house, leaving 23% to be covered by an auxiliary source.

The economic study is repeated for the 65 solutions using the international minimum benchmark price, which is equal to $ 1.05 /liter. The highest NPV is found to be \$411,341 at solution number 60, where A_{SC}=131.1 m^2 and V_{tank}=12 m^3. The investment cost of this SHS design is \$95,889 and it covers 90% of the heat demand of the house, leaving 10% to be covered by an auxiliary source. It has to be noted that increasing the fuel price to match the minimum international value, which is almost 4 times the local price, caused a tremendous increase in the NPV. The highest NPV using the international fuel price is 21 times the highest NPV calculated using the local prices.

Thus, it can be concluded that if the fuel prices in Egypt increase to reach the minimum international fuel price, then heating the modelled poultry house using the SHS will be vital and extremely profitable in comparison to the conventional fuel-based heating system. Yet, given the current local fuel price, it is still economically acceptable to use the SHS for heating the modelled house when selecting the suitable design.

The study is extended to evaluate the effect of the house's infiltration rate on the performance of the SHS. The modelled house under study is estimated to have an infiltration rate of 1.19 ACH, which is a very high value compared to the standard ASHRAE value of 0.45 ACH for a tight building. The value of infiltration rate of the modeled house is varied to be 0.45 ACH and 2 ACH to check the effect of increasing and decreasing its value. The economic study is repeated using these 2 values for 3 different designs. The 3 designs are selected from the solution space to be solution 12, 29 and 46 to cover different regions of the solution space and to include the best economic solution, which is solution 46. The NPV for each solution is found to increase with decreasing the infiltration rate. For the best economic design, solution 46, the NPV can increase by a value of \$7600 when reducing the infiltration rate from 1.19 to 0.45 ACH, where the auxiliary energy needed drops from 23% to 17%. Enhancing the infiltration rate to be 0.45, will increase the fuel savings for the first year by a value of \$1000. Therefore, it is recommended to enhance the tightness of the building to reach a lower infiltration rate for better economic results.

A further study is done to check the effect of using latent heat storage technique on the performance of the SHS. The water storage tank used in the SHS is replaced by a water tank with embedded PCM modules. The new tank is modelled using TYPE 840 TRNSYS component, which is developed by the IEA SHC TASK 32 and is obtained from one of the authors, Dr. Hermann Schranzhofer. The PCM used is Sodium Acetate Trihydrate with embedded graphite of melting temperature between 56 and 60 °C. The NPV calculated for the SHS with embedded PCM is found to be less than that of the SHS with sensible heat storage by a value of $3,627. Therefore, given the cost of the PCM modules, using the described latent heat storage technique yielded a negative effect on the economic performance of the system. However, it has to be noted that the fan coil unit performance is rated at inlet water temperature of 90°C, which is much higher than the phase change temperature of the material used.

Finally, in an attempt to build a completely green heating solution for poultry houses, bio-digesters are introduced to produce biogas from the waste/manure of the birds. From the calculations, it is found that 12 units of a simple anaerobic bio-digester model called Deenbandhu of capacity 6 m^3 each are capable of covering the auxiliary energy needed by the SHS design of solution 46. The total cost for establishing the 12 units as per SKG Sangha foundation in Egypt is found to be $16800. An economic study is carried out for the all-green heating system, where the bio-digesters are used as the auxiliary source for the SHS design of solution 46. The NPV value is calculated to be $ 30,237, which is much higher than that of using diesel fuel as the auxiliary source of energy. The 6 m^3 of biogas produced daily from the bio-digesters are stored to be used as the auxiliary source of energy for the SHS all over the year. Therefore, it can be concluded that the modelled poultry house produces enough manure that can be transformed into biogas using the bio-digesters to complement the SHS in covering the full heating demand of the house economically. However, it has to be noted that the bio-digesters require an area of 254 m^2 to be established. Also, the complexity of mixing the waste with water before entering the bio-digester and handling the excess discharge of the bio-digester has to be put into consideration when applying the suggested all-green heating solution.

QUESTIONS

1. Calculate the required heating load for a 500 m^2 broiler house for 5000 chicks using your country's weather data.

2. Compare between the cost of heating the above broiler house (in question 1) using the following traditional energy sources:
   a. Electricity
   b. LPG
   c. Natural Gas
   d. Fuel oil (Diesel fuel)
3. Calculate the cost of heating the broiler house discussed in question 1 using solar water heating system, showing the capital cost and running cost.
4. Compare between heating the broiler house discussed in question 1 using normal solar water heating system and solar water heating system with integrated phase change material (PCM).
5. Compare between heating the broiler house discussed in question 1 using solar water heating system integrated with 2 different types of PCM:
   a. Paraffin wax with melting temperature of about 60°C
   b. Sodium Acetate Trihydrate with embedded graphite
6. Select the optimum energy mix combination of the traditional energy sources discussed in question 2 and the normal solar water heating system to heat the broiler house discussed in question 1.
7. Compare between the solar water heating system to be used for heating a 500 m^2 broiler house occupied with 5000 chicks *versus* the same house occupied by 10000 chicks.
8. Evaluate the environmental impact of using solar water heating system *versus* using fuel oil heating in terms of CO_2 emissions reduction. How would a carbon credit scheme impact the economics of heating the broiler house?

REFERENCES

[1] J.O. Donald, *Environmental Management in the Broiler House..* Aviagen, 2009.

[2] M. Kharseh, and B. Nordell, "Sustainable Heating and Cooling Systems for Agriculture", *Int. J. Energy Res.,* vol. 35, no. 5, pp. 415-422, 2011. [http://dx.doi.org/10.1002/er.1699]

[3] J.P. Lesschen, M. Van den Berg, H.J. Westhoek, H.P. Witzke, and O. Oenema, "Greenhouse Gas Emission Profiles of European Livestock Sectors", *Anim. Feed Sci. Technol.,* vol. 166, pp. 16-28, 2011. [http://dx.doi.org/10.1016/j.anifeedsci.2011.04.058]

[4] A. Atilgan, and A. Akyuz, "The Investigation of Heating and Cooling Days with the Method of Degree-day in Broiler Poultry Housing", *Asian J. Anim. Vet. Adv.,* vol. 2, no. 3, 2007.

[5] R.S. Gates, D.G. Overhults, and S.H. Zhang, "Minimum ventilation for modern broiler facilities", *Trans. ASAE,* vol. 39, no. 3, pp. 1135-1144, 1996. [http://dx.doi.org/10.13031/2013.27605]

[6] M. Thirugnanasambandam, S. Iniyan, and R. Goic, "A Review of Solar Thermal Technologies", *Renew. Sustain. Energy Rev.,* vol. 14, no. 1, pp. 312-322, 2010. [http://dx.doi.org/10.1016/j.rser.2009.07.014]

[7] S.D. Sharma, and K. Sagara, "Latent Heat Storage Materials and Systems: A Review", *Int. J. Green Energy,* vol. 2, no. 1, pp. 1-56, 2005. [http://dx.doi.org/10.1081/GE-200051299]

[8] A. Sharma, V.V. Tyagi, C.R. Chen, and D. Buddhi, "Review on thermal energy storage with phase change materials and applications", *Renew. Sustain. Energy Rev.,* vol. 13, no. 2, pp. 318-345, 2009. [http://dx.doi.org/10.1016/j.rser.2007.10.005]

[9] V. Torretta, E.C. Rada, I.A. Istrate, and M. Ragazzi, "“Poultry Manure Gasification and its Energy Yield,” UPB Sci. Bull. Ser", *Mech. Eng.,* vol. 75, pp. 231-238, 2013.

[10] S. Sakar, K. Yetilmezsoy, and E. Kocak, "Anaerobic digestion technology in poultry and livestock waste treatment--a literature review", *Waste Manag. Res.,* vol. 27, no. 1, pp. 3-18, 2009. [http://dx.doi.org/10.1177/0734242X07079060] [PMID: 19220987]

[11] TRNSYS, *Available at: www.trnsys.com.*

[12] Aviagan, *Broiler Management Guide.* 2009.

[13] Google Sketchup, *Google, Inc.*

[14] FACCO, *Poultry Equipments..*

[15] ASHRAE, *2001 ASHRAE Handbook Fundamentals.* American Society of Heating, Refrigerating, and Air-Conditioning Engineers, 2001.

[16] J. Donald, *Tests Show Fan Shutter Air Leakage Causes Cold Weather Problems..* Alabama Cooperative Extension System: Auburn, 1997.

[17] C.M. Wathes, "Air Hygiene", In: *Measuring and Auditing Broiler Welfare..* CABI Publishing, 2004, pp. 117-131. [http://dx.doi.org/10.1079/9780851998053.0117]

[18] Aviagen, *Arbor Acres Broiler Performance Objectives.* 2007.

[19] A. Butterworth, "Infectious Disease: Morbidity and Mortality", In: *Measuring and Auditing Broiler Welfare..* CABI Publishing, 2004, pp. 61-70. [http://dx.doi.org/10.1079/9780851998053.0061]

[20] Farm Manager, *Mortality Rate of Al Menia House.* 2013.

[21] Aviagen, *Arbor Acres Broiler Management Guide, .*

[22] Solar Energy Laboratory Univ, *Arbor Acres Broiler Management Guide, .*

[23] *Interview with Dr. Khaled Mostafa the General Manager of Grand Parent Sector at Cairo Poultry Group, .*

[24] Egyptian Cabinet, *Modifying Fuel Prices..* Egyptian Official Gazette: Egypt, 2014.

[25] GIZ Deutsche Gesellschaft für Internationale Zusammenarbeit GmbH, *International Fuel Prices 2012 / 2013,* .

[26] J.A. Duffie, and W.A. Beckman, *Solar Engineering of Thermal Processes..* 4th ed John Wiley & Sons, Inc., 2013.
[http://dx.doi.org/10.1002/9781118671603]

[27] J.A. White, K.E. Case, D.B. Pratt, and M.H. Agee, *Principles of Engineering Economic Analysis..* 4th ed John Wiley & Sons, Inc., 1998.

[28] Apricus Solar Hot Water: Evacuated Tubes, *Apricus Solar Co. Ltd.,* 2014. [Online]. Available at: http://www.apricus.com/html/evacuated_tubes.htm#.VIugRiuUf4s. [Accessed: 01-Dec-2014]

[29] İ. Dinçer, and C. Zamfirescu, *Sustainable energy systems and applications..* Springer, 2011.

[30] Central Bank of Egypt, [Online]. Available at: http://www.cbe.org.eg/. [Accessed: 07-Dec-2014]

[31] *Egypt's Electricity Tariff Law..* Ministry of Finance: Egypt, 2014, p. 7.

[32] M. El-Fiqi, *Energy subsidies in the balance.* Al Ahram Weekly: Cairo, 2011.

[33] *Income Tax Law..* Ministry of Finance: Egypt, 2005.

[34] J. Ondraczek, N. Komendantova, and A. Patt, "WACC the dog: The effect of financing costs on the levelized cost of solar PV power", *Renew. Energy,* vol. 75, pp. 888-898, 2015.
[http://dx.doi.org/10.1016/j.renene.2014.10.053]

[35] Egyptian Ministry of Finance, [Online]. Available at: http://www.mof.gov.eg/. [Accessed: 07-De--2014]

[36] D.C. Montgomery, *Design and Analysis of Experiments, Eighth..* John Wiley & Sons, Inc., 2013.

[37] American Society of Heating Refrigerating and Air-Conditioning Engineers, *ASHRAE Handbook: Fundamentals..* ASHRAE: Atlanta, GA, 2001.

[38] W. Streicher, J. Bony, S. Citherlet, A. Heinz, P. Puschnig, H. Schranzhofer, and J.M. Schultz, *Simulation Models of PCM Storage Units..* Graz, 2008.

[39] H. Schranzhofer, *Personal Communication,* . hermann.schranzhofer@tugraz.at

[40] European Committee for Standardization CEN, *Heating systems in buildings — Method for calculation of system energy requirements and system efficiencies — Part 4-3 Space heating generation systems, thermal solar systems,* .

[41] P. Puschnig, *Description of a Storage Model for Phase Change Materials,* .

[42] H. Schranzhofer, P. Puschnig, A. Heinz, and W. Streicher, "Validation of a TRNSYS Simulation Model For PCM Energy Storages and PCM Wall Construction Elements", *Ecostock Conference,* 2006pp. 2-7

[43] Phase Change Material Products Limited, *PlusICE Phase Change Materials, Webpage,* 2014. [Online]. Available at: www.pcmproducts.net. [Accessed: 22-Nov-2014]

[44] M. D. Heidari, and M. Omid, *Energy,* vol. 36, no. 11, pp. 6536-6541, 2011.
[http://dx.doi.org/10.1016/j.energy.2011.09.011]

[45] S.M. ElHaggar, *Solid waste management: alternatives - innovations - solutions.*. Dar Elfikr Elarabi, 2004.

[46] K.K. Kudaravalli, *Education Manual.*. Kolar: Karnataka, India, 2014.

[47] A. Medhat, *SKG Sangha Foundation in Egypt.*.

SUBJECT INDEX

Salah El-Haggar, Sarah Mousa & Mohamed El-Morsi (Eds.)
All rights reserved-© 2016 Bentham Science Publishers

T

U

V

W

X

Z

Made in United States
Orlando, FL
05 November 2026

90200368R10150